Quantum Mechanics

Trilochan Pradhan

Honorary Professor Emeritus
Institute of Physics
Bhubaneswar

Universities Press

QUANTUM MECHANICS

First published in the UK in 2009 by

Anshan Ltd
11 A Little Mount Sion
Tunbridge Wells
Kent. TN1 1YS
UK
Tel: +44 (0) 1892 557767
Fax: +44 (0) 1892 530358
e-mail: info@anshan.co.uk
Web Site: www.anshan.co.uk

Sold and distributed in all countries worldwide
except India, Pakistan, Sri Lanka, Bangladesh, Nepal, Bhutan and the Maldives.

British Library Cataloguing in Publication Data
A catalogue record for this book is available from the British Library

ISBN 978 1848290 389

Original Indian Publisher
Universities Press (India) Private Limited
Registered Office
3-6-747/1/A & 3-6-754/1 Himayatnagar, Hyderabad 500 029 (AP.) India
email: info@universitiespress.com

Distributed in India, Pakistan, Sri Lanka, Bangladesh, Nepal, Bhutan and the Maldives by
Orient Blackswan Private Limited
3-6-752 Himayatnagar, Hyderabad 500 029 (AP.) India

Typeset by
Krishtel eMaging Solutions Private Limited, Chennai 600 017, India

Printed at
Graphica Printers, Hyderabad 500 013, India

Published by
Universities Press (India) Private Limited
3-6-747/1/A& 3-6-754/1 Himayatnagar, Hyderabad 500 029 (AP.) India
email: info@universitiespress.com

To my wife, Sanjukta,
son and daughter, Ashok and Asima,
their spouses Jhansi and Prasanta,
and grandson Pradeepta (Apu)

Contents

Preface

I have had the privilege of being taught quantum mechanics by Enrico Fermi at the University of Chicago in 1953–54—a subject which I later taught for nearly 40 years to M.Sc., post-M.Sc. and pre-doctoral students at the Ravenshaw College, Cuttack, Saha Institute of Nuclear Physics, Calcutta and Institute of Physics, Bhubaneswar, This book is the outcome of the urge I felt to consolidate my lectures—to present many topics that are not covered by most books of today nor classic texts of yester years. These include:

- the genesis—the way Heisenberg and Dirac invoked Bohr's correspondence principle (which Sommerfeld termed as his magic wand) to discover quantum mechanics; the way de Broglie arrived at his famous matter–wave formula and Schrödinger discovered the mechanics of these waves, viz. wave-mechanics,
- the first quantized theory of photons and neutrinos—most books give the second quantized theory,
- Bohr-Sommerfeld 'action' as differential operators with their eigenvalues n and l and their corresponding eigenfunctions,
- parabose and parafermi symmetries of identical particles,
- Dirac's initiation of Lagrangian formulation of quantum mechanics, also known as transformation theory, and its elaboration and completion by Feynman in which the probability of path takes the place of the probability of position of the Hamiltonian formulation,
- topological phase of the wavefunction in Bohm–Aharonov, Aharonov–Casher and neutron interferometer experiments which are examples of the Berry phase,
- Schrödinger's cat, EPR paradoxes and Bell's inequality and
- quantum beats such as Stark and exchange oscillations similar to $K^o\overline{K}^o$ and neutrino oscillations in particle physics.

The book is limited to non-relativistic theory. All the topics are presented in a concise and precise manner. Although it is not addressed to beginners, it is not necessary to have any prior acquaintance with the subject because what has been presented is self-contained. Applications are restricted to atoms; nuclei and solids are not covered. Most chapters have problems at the end, quite a few of which were assigned by Fermi to his 1953–54 class. These problems are such that the basic principles become clear after solving them; hints have been provided wherever possible.

I have had fruitful discussions with professors Avinash Kare of the Institute of Physics, Bhubaneswar and Prasanta Panigrahi (PP) of the Physical Research Laboratory, Ahmedabad on several topics covered in this book, for which I am grateful. PP also went through the manuscript with a critical eye. I have been assisted by K. Srinivas of the School of Physics, University of Hyderabad, in getting the manuscript done in LaTeX and by R. K. Patra of the Institute of Physics, Bhubaneswar, in carrying out the corrections in it. I am thankful to both for their work.

September 2007

Trilochan Pradhan
Institute of Physics
Bhubaneswar, India

1 Genesis

1.1 Introduction

Quantum mechanics was discovered by Heisenberg and Dirac through Niels Bohr's correspondence principle, and wave mechanics was formulated by Schrödinger, who obtained the differential equation for matter waves postulated by Louis de Broglie; Max Born provided the necessary physical interpretation for the same. It was later found that these two independent and parallel developments are equivalent. It is worthwhile to present these two developments first before proceeding to the foundations and applications of the subject. We begin with a presentation of Heisenberg's use of correspondence principle to discover the non-commutative nature of position and momentum of particles, followed by a discussion of his physical insight into this with a 'thought' experiment and the enunciation of the uncertainty principle. We then present the logical development of de Broglie's matter wave postulate, how Schrödinger obtained differential equation for these waves and Max Born's interpretation of the wavefunction.

1.2 Correspondence Principle

In the theories of Max Planck and Niels Bohr, quantum postulates were superimposed on the classical theory. While in Planck's theory, it was assumed that the frequency of radiation equals the oscillation frequency,

$$\nu_{\text{radiation}} = \nu_{\text{oscillator}}, \tag{1.1}$$

in Bohr's theory, the radiation frequency ν was shown to be equal to the classical orbital frequency f and the harmonics thereof for large principal quantum numbers.

$$\nu = \Delta n f, \Delta n = n_1 - n_2 \tag{1.2}$$

for $n \gg \Delta n, n = \frac{1}{2}(n_1 + n_2)$.

Equalities (1.1) and (1.2) highlight the correspondence between classical and quantum frequencies. The second of these is a particular case of

Bohr's correspondence principle. Its derivation can be obtained from Bohr's quantization formula:

$$S = \oint pdq = nh, \tag{1.3}$$

from which we obtain

$$\Delta S = \Delta nh. \tag{1.4}$$

From the definition

$$\nu = \frac{\Delta E}{h} \quad \text{and} \quad f = \frac{dE}{dS}, \tag{1.5}$$

it then follows that

$$\nu = \left(\frac{\Delta E}{\Delta S}\right)\Delta n. \tag{1.6}$$

This leads to the correspondence principle (1.2) if the classical differential quotient $\frac{dE}{dS}$ equals the difference quotitient $(\Delta E/\Delta S)$. From the plots of E against S given in Fig. 1.1 for the hydrogen atom and harmonic oscillator, it will be seen that while for the former $\frac{\Delta E}{\Delta S} = \frac{dE}{dS}$ for large S only, for the latter it is true for all S.

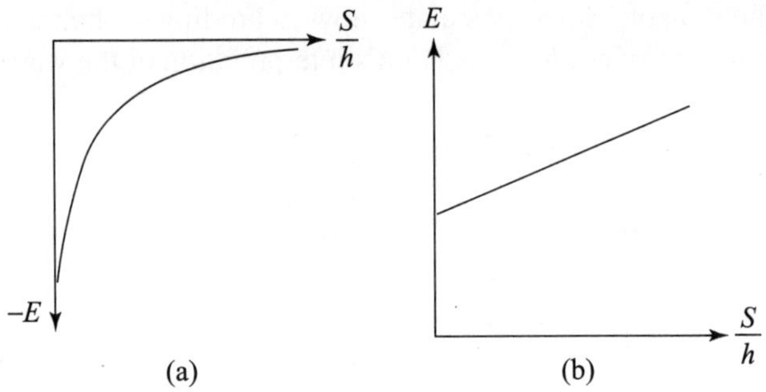

Fig. 1.1 *Plots of E vs. S for (a) hydrogen atom, (b) harmonic oscillator.*

1.3 Birth of Quantum Mechanics

Quantum mechanics was discovered by Heisenberg and Dirac following parallel paths through the use of correspondence principle. Let us start with Heisenberg's approach. He proposed that the position x of a particle appearing in its equation of motion

$$\ddot{x} + f(x) = 0 \tag{1.7}$$

whose Fourier series in classical mechanics is written as

$$x = \sum_{\tau=-\infty}^{\infty} a_\tau e^{i\tau\omega t}, \tag{1.8}$$

be replaced by

$$x = \sum_{\tau=-\infty}^{\infty} a(n, n-\tau) e^{i\omega(n,n-\tau)\tau t} \tag{1.9}$$

in quantum mechanics, where the amplitude $a(n, n-\tau)$ and frequency $\omega(n, n-\tau)$ represent two states rather than one. The correspondence between classical amplitude a_τ and quantum amplitude $a(n, n-\tau)$ is that the latter varies slowly with n for large n and small τ, i.e.,

$$\lim_{n\gg\tau} a(n, n-\tau) \longrightarrow a_\tau. \tag{1.10}$$

This can be used in Bohr's quantization condition

$$S = \oint p dq = \int_0^T dt p\dot{q} = nh. \tag{1.11}$$

Using classical theory series

$$p = \sum_{\tau=-\infty}^{\infty} p_\tau e^{i\omega\tau t}; \quad q = \sum_{\tau=-\infty}^{\infty} q_\tau e^{i\omega\tau t}, \tag{1.12}$$

equation (1.11) can be written as

$$2\pi i \sum_{\tau=-\infty}^{\infty} \tau \frac{\partial}{\partial S}(q_\tau p_{-\tau}) = 1. \tag{1.13}$$

In this equation we now use the correspondence principle according to which the differential quotient is to be replaced by difference quotient in order to go over from classical mechanics to quantum mechanics. This gives

$$\sum_{\tau=-\infty}^{\infty} [q(n+\tau, n)p(n, n+\tau) - q(n, n-\tau)p(n-\tau, n)] = \frac{h}{2\pi i}. \tag{1.14}$$

In this relation, we now make use of $p = m\dot{q}$ and write

$$p(n, n-\tau) = p(n, k).$$

Setting the $p(n, m)$ and $q(n, m)$ with negative indices equal to zero, we finally obtain

$$\sum_k [p(n,k)q(k,n) - q(n,k)p(k,n)] = \frac{h}{2\pi i}. \tag{1.15}$$

This relation was first obtained by Heisenberg and later generalised jointly by Born, Heisenberg and Jordan.

Dirac accepted Heisenberg's modification of classical amplitude and proceeded to find out the classical equivalent of commutator of quantum variables $x(n, n-\alpha)$ and $y(n, n-\alpha)$ which runs as follows:

$$\begin{aligned}
x(n, n-\alpha)&y(n-\alpha, n-\alpha-\beta) - y(n, n-\beta)x(n-\beta, n-\alpha-\beta) \\
&= [x(n, n-\alpha) - x(n-\beta, n-\beta-\alpha)]y(n-\alpha, n-\alpha-\beta) \\
&\quad - [y(n, n-\beta) - y(n-\alpha, n-\alpha-\beta)]x(n-\beta, n-\alpha-\beta) \\
&= h\sum_r \left[\beta_r \frac{\partial x_{\alpha k}}{\partial k_r} y_{\beta k} - \alpha_r \frac{\partial y_{\beta k}}{\partial k_r} x_{\alpha k}\right] \\
&= \frac{ih}{2\pi} \sum_{\alpha+\beta=n-m} \sum_r \left[\frac{\partial}{\partial k_r}\left(x_r e^{i\alpha_s\omega_s t}\right) \frac{\partial}{\partial \omega_r}\left(y_\beta e^{i\beta_s\omega_s t}\right)\right. \\
&\quad \left. - \frac{\partial}{\partial k_r}\left(y_r e^{i\beta_s\omega_s t}\right) \frac{\partial}{\partial \omega_r}\left(x_\beta e^{i\alpha_s\omega_s t}\right)\right],
\end{aligned} \tag{1.16}$$

where $x_{\alpha k}$ stands for the correspondence limit of $x(n, n-\alpha)$ and $k_r = n_r h$ or $(n_r + \alpha_r)h$, these being practically equivalent.

It will be noticed that (1.16) is equivalent to

$$xy - yx = -\frac{ih}{2\pi}\sum_r \left(\frac{\partial x}{\partial S_r}\frac{\partial y}{\partial \theta_r} - \frac{\partial y}{\partial S_r}\frac{\partial x}{\partial \theta_r}\right), \tag{1.17}$$

where $S_r = \kappa_r$ is the action variable and $\theta_r = \frac{\omega_r t}{2\pi}$ is the angle variable. Identity (1.17) can be written as

$$xy - yx = \frac{i\hbar}{2\pi}\{x, y\}_{\text{PB}}, \tag{1.18}$$

i.e., the quantum commutator is $\frac{ih}{2\pi}$ times the classical Poisson bracket (PB). If x and y are taken as p and q, then (1.18) gives the Born–Heisenberg–Jordan relation (1.15).

1.4 Heisenberg's Uncertainty Principle

The position and momentum relation obtained by Heisenberg which is so different from that of classical mechanics, led him to examine through a thought

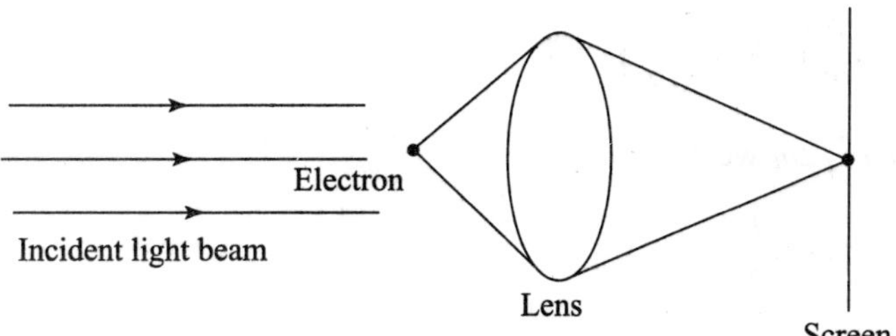

Fig. 1.2 *Heisenberg's thought experiment*

experiment what these variables actually represent. Accurate measurement of position of an electron can be done by shining light on it, which after getting scattered, enters a microscope's lens to give an image of it. This is schematically illustrated in the Fig. 1.2

It is known from optics that the location of the position of the image is limited by the resolving power of the lens to

$$\Delta x = \frac{\lambda}{\sin\theta}, \tag{1.19}$$

where θ is the aperture of the lens. In order to minimise the effect of the radiation pressure of light, one makes its intensity as low as possible—so weak that only one light quantum gets scattered by the electron. The collision of the photon imparts to the electron a momentum with a range

$$\Delta p \sim p\sin\theta = \frac{\hbar}{\lambda}\sin\theta, \tag{1.20}$$

where p is the momentum of the photon. The electron momentum is thus uncertain to the extent given by (1.20). From (1.19) and (1.20) we get

$$\Delta x \Delta p \sim h, \tag{1.21}$$

which is the uncertainty principle. If we want to measure position more accurately, we need to minimise Δx, which, according to (1.19), can be done by making λ small. But then according to (1.20), this increases uncertainty in momentum, Δp, the product of the two uncertainties remaining constant. This thought experiment underlines the dynamics of quantum systems which is mathematically expressed by the Heisenberg–Dirac commutation relation.

We remark in passing that there is an uncertainty relation between energy and time

$$\Delta E \cdot \Delta t \sim h, \tag{1.22}$$

which is sometimes called the Heisenberg–Bohr uncertainty relation. This can be obtained from the position–momentum relation by the following manipulations. From (1.21) we get

$$(\Delta p)\left(\frac{p}{m}\right)(\Delta x)\left(\frac{m}{p}\right) \sim h. \tag{1.23}$$

From $E = p^2/2m$ we get

$$\Delta E = (\Delta p)\frac{p}{m}. \tag{1.24}$$

Further,

$$(\Delta x)\left(\frac{m}{p}\right) = \frac{\Delta x}{v} = \Delta t. \tag{1.25}$$

Substitution of (1.24) and (1.25) in (1.23) gives the desired relation (1.22). It may be noted that this derivation is valid for free particles.

1.5 de Broglie's Matter Waves

The fact that light waves behave as particles in photoelectric effect whose energy is related to the frequency of these waves by Einstein's quantum relation $E = h\nu$, and that electrons in atoms have discrete energies as Bohr showed and which is similar to the discrete frequencies of waves on strings and membranes in classical physics led de Broglie to speculate that material particles like electrons are somehow associated with waves. He noted that if this were the case, there would be unification of matter and light and as in photo-effect wavelength; λ of light would be related to its momentum p by $\lambda = h/p$, the waves associated with material particles would also have the same relation between the wavelengths and their momenta.

The derivation of his relation, as put forth by him, runs as follows. For a particle at rest, the energy–frequency relation could be written as

$$mc^2 = h\nu, \tag{1.26}$$

and the wave associated with it could be represented by

$$\psi = a \sin 2\pi\nu t.$$

In a moving frame this relation would have the form

$$\psi = a \sin 2\pi\nu\left(\frac{t' - x'v/c^2}{\sqrt{1 - v^2/c^2}}\right). \tag{1.27}$$

When compared with a travelling wave

$$\psi = a \sin 2\pi\left(\nu t' - \frac{x'}{\lambda}\right), \tag{1.28}$$

it gives

$$\frac{2\pi\nu v}{c^2\sqrt{1-v^2/c^2}} = \frac{2\pi}{\lambda}, \tag{1.29}$$

which on account of (1.26) gives

$$\lambda = \frac{h}{mv}\sqrt{(1-v^2/c^2)} = \frac{h}{p}. \tag{1.30}$$

1.6 Differential Equation for Matter Waves

If we represent waves associated with a material particle moving in a time-independent potential by the function $\psi(x)$ which in general can be written as a Fourier integral

$$\psi(\vec{x},t) = \frac{1}{(2\pi)^3}\int d^3k e^{i\vec{k}\cdot\vec{x}-i\omega t}\phi(\vec{k},\omega), \tag{1.31}$$

and use the energy–frequency relation

$$\omega = \frac{E}{\hbar} = \left(\frac{p^2}{2m}+V\right)\frac{1}{\hbar} \tag{1.32}$$

as well as the de-Broglie relation

$$p = \hbar k = \frac{\hbar}{\lambda\!\!\!^{-}}, \quad \lambda\!\!\!^{-} = \frac{\lambda}{2\pi} \tag{1.33}$$

in the integral, we get

$$\psi(\vec{x},t) = \frac{1}{(2\pi)^3}\int d^3k e^{i\vec{k}\cdot\vec{x}-i\left(\frac{k^2\hbar^2}{2m}+V\right)t}\phi(\vec{k},\omega). \tag{1.34}$$

Further, differentiation of (1.31) gives

$$-\frac{\hbar^2}{2m}\nabla^2\psi(\vec{x},t) = \frac{1}{(2\pi)^3}\int d^3k\frac{k^2\hbar^2}{2m}\phi(\vec{k},\omega)e^{i\vec{k}\cdot\vec{x}-i\left(\frac{k^2\hbar^2}{2m}+V\right)t}, \tag{1.35}$$

$$i\hbar\frac{\partial\psi(\vec{x},t)}{\partial t} = \frac{1}{(2\pi)^3}\int d^3k\left(\frac{k^2\hbar^2}{2m}+V\right)\phi(\vec{k},\omega)e^{i\vec{k}\cdot\vec{x}-i\left(\frac{k^2\hbar^2}{2m}+V\right)t}. \tag{1.36}$$

A comparison of these two identities leads to the differential equation

$$i\hbar\frac{\partial\psi}{\partial t} = -\frac{\hbar^2}{2m}\nabla^2\psi + V\psi, \tag{1.37}$$

which is the differential equation for matter waves obtained by Schrödinger.*

---

*See Appendix for Schrödinger's own derivation.

The physical interpretation of the wave function $\psi(x)$ was provided by Max Born, who put forth the view that it is a complex function and its square modulus represents the probability of finding the particle which it is associated with. This interpretation is similar to the case of light waves, where the intensity of light as calculated from wave theory is proportional to the number of photons given by quantum theory.

It may be seen from (1.32) and (1.37) that

$$i\hbar\frac{\partial\psi}{\partial t} = E\psi, \tag{1.38}$$

and from (1.31) that

$$-i\hbar\vec{\nabla}\psi = \vec{p}\psi\,. \tag{1.39}$$

These two equations tell us that if energy E and momentum $\vec{p}$ of classical mechanics are replaced by differential operators

$$E = i\hbar\frac{\partial}{\partial t},\ \vec{p} = -i\hbar\vec{\nabla} \tag{1.40}$$

in the classical equation of motion, one obtains the wave mechanical equation of motion. These two relations are known as Schrödinger's quantization prescription.

Finally, we note that on using this prescription we have

$$\begin{aligned}(p_i x_j - x_j p_i)\psi &= -i\hbar[\partial_i(x_j\psi) - x_j\partial_i\psi] \\ &= -i\hbar\delta_{ij}\psi,\end{aligned} \tag{1.41}$$

which is Heisenberg–Dirac commutation relation. We thus see that the quantum mechanics of Heisenberg and Dirac, and wave mechanics of Schrödinger are equivalent.

Problems

1. The polarisability α is given

$$\vec{P} = \alpha\vec{E} = e\vec{x}$$

where $\vec{E}$ an electric field and

$$\frac{d^2\vec{x}}{dt^2} + \omega_0^2\vec{x} = \frac{e}{m}\vec{E}.$$

Taking

$$\vec{x} = \sum_{n=-\infty}^{\infty} \vec{a}(n, n-\beta) e^{i\omega(n,n-\beta)},$$

obtain an expression for α in terms of

$$f(n+\beta, n) = \frac{m}{\hbar} |\vec{a}(n, n+\beta)|^2 \, \omega(n, n+\beta),$$

$$f(n, n-\beta) = \frac{m}{\hbar} |\vec{a}(n, n-\beta)|^2 \, \omega(n, n+\beta).$$

The result is Kramers–Heisenberg dispersion formula.

2. Use

$$x = \sum_{n=-\infty}^{\infty} a(n, n-\beta) e^{i\omega(n,n-\beta)}$$

in Bohr's quantization rule

$$m \oint \dot{x} dx = nh$$

to obtain the sum rule for the coefficients a. The sum rule so obtained is the Thomas–Kuhn sum rule.

Bibliography

[1] N. Bohr, *Kgl. Danske Vid. Selsk. Skr., nat. math. Afd.*, **8**, Raekke IV 1 (1918).
[2] W. Heisenberg, *Z. Phys.*, **33**, 879 (1925).
[3] P.A.M. Dirac, *Proc. Roy. Soc.*, A**109**, 642 (1925).
[4] W. Heisenberg, *Proc. Roy, Soc.*, A**43**, 172 (1927).
[5] M. Born, and P. Jordan, *Z. Phys.*, **34**, 858 (1925).
[6] L. de Broglie, *Ann. de Phys.*, **10**, 3, 22 (1925).
[7] E. Schrödinger, *Ann. Physik*, **79**, 361 (1925); **80**, 437 (1926); **81**, 109 (1926).
[8] M. Born, *Z. Physik*, **38**, 803 (1926).
[9] B.L. Van Der Waerden, Editor, *Sources of Quantum Mechanics*, Dover Publications Inc., New York (1968).

2 Foundations

2.1 Ground Rules of Quantum Mechanics

In the previous chapter we have seen that the inadequacy of classical mechanics for the description of phenomena at the atomic level led to the birth of a new mechanics, wherein position and momentum of a particle do not commute and cannot be determined simultaneously with absolute accuracy. It was also seen that energy and momentum are to be represented by differential operators which operate on wave functions. In this chapter we shall present the ground rules (axioms) of this new theory. These are:

1. The state of a system is described by a wave function $\psi(\vec{x})$, which in general is complex and normalisable, i.e.,

$$P = \int d\vec{x}\psi^*(\vec{x})\psi(\vec{x}) = \text{finite}. \tag{2.1}$$

2. P, the total probability, is conserved in time as a consequence of the continuity equation,

$$\frac{\partial\rho}{\partial t} + \vec{\nabla}\cdot\vec{j} = 0,$$

where $\rho = \psi^*\psi$ and $\vec{j} = \frac{\hbar}{2im}[\psi^*(\vec{\nabla}\psi) - (\vec{\nabla}\psi^*)\psi]$ and which follows from the Schrödinger equation

$$i\hbar\frac{\partial\psi}{\partial t} = -\frac{\hbar^2}{2m}\nabla^2\psi + V\psi.$$

3. Observables (physical quantities) of the system, which have to be real, are represented by operators Ω which are hermitian, i.e.,

$$\int d\vec{x}\psi^*(\vec{x})\Omega\psi(\vec{x}) = \int d\vec{x}(\Omega\psi(\vec{x}))^*\psi(\vec{x}). \tag{2.2}$$

4. The measurement of an observable represented by operator Ω in a state $|\psi_n\rangle$ with wave function $\psi_n(x)$ gives the eigenvalue λ_n,

$$\Omega|\psi_n\rangle = \lambda_n|\psi_n\rangle\,. \tag{2.3}$$

Here it is implied that conditions are created such that the eigenvalue appears as a result of the measurement.

According to the Copenhagen interpretation, before the measurement the system is in several states. Measurement forces the system to collapse to a given state $|\psi_n\rangle$. In other words, the quantum systems have no properties until they are measured.

5. The eigenfunctions of a hermitian operator form an orthonormal complete set in the Hilbert space of states:

$$\langle\psi_n(\vec{x})|\psi_m(\vec{x}')\rangle = \int d\vec{x}\,\psi_n^*(\vec{x})\psi_m(\vec{x}) = \delta_{n,m},$$

$$\sum_n |\psi_n(\vec{x})\rangle\langle\psi_n(\vec{x}')| = \delta(\vec{x}-\vec{x}'), \tag{2.4}$$

where $\delta(\vec{x})$ is Dirac delta function.

This is valid for non-degenerate states. If there are degenerate states i.e., more than one state having the same energy but different eigenvalues for some other observable, one has to construct linear combinations of these eigenfunctions which are orthogonal. This procedure is known as Schmidt orthogonalisation.

Degeneracy occurs when two conserved operators F and G do not commute:

$$[F, H] = 0, [G, H] = 0 \tag{2.5}$$

$$[F, G] \neq 0, \tag{2.6}$$

so that,

$$F\psi = f\psi, \quad \text{and} \quad H\psi = \epsilon\psi. \tag{2.7}$$

In that case

$$G\psi \neq g\psi, \tag{2.8}$$

where g is a c number, on account of (2.6). On the other hand, on account of (2.5),

$$H(G\psi) = G(H\psi) = \epsilon(G\psi). \tag{2.9}$$

From (2.7) and (2.9) it then follows that both ψ and $G\psi$, which are not same, have the same energy. Thus there is degeneracy.

6. Any normalisable wave function $\psi(x)$ can be expanded in terms of a complete set of orthonormal eigenfunctions $u_n(x)$:

$$\psi(x) = \sum_n c_n u_n(x), \tag{2.10}$$

where

$$c_n = \langle u_n(x)|\psi(x)\rangle, \tag{2.11}$$

and $|c_n|^2$ represents the probability of occurrence of the eigenstate $|u_n(x)\rangle$ with eigenvalue λ_n and hence,

$$\sum_n |c_n|^2 = 1. \tag{2.12}$$

7. The expectation (average) value of $\bar{\Omega}$ an operator Ω for the system represented by wave function $\psi(x)$ is given by

$$\bar{\Omega} = \frac{\langle\psi|\Omega|\psi\rangle}{\langle\psi|\psi\rangle}. \tag{2.13}$$

8. Since, on account of uncertainty principle, an observable which has a definite value at one instant of time, may not have a definite value at a subsequent instant of time, differentiation with respect to time cannot be defined like in classical mechanics. Therefore, the time derivative $\dot{\Omega}$ of an operator Ω is defined as that operator whose expectation value equals the time derivative of the expectation value of the operator Ω:

$$\langle\dot{\Omega}\rangle = \frac{d}{dt}\langle\Omega\rangle. \tag{2.14}$$

Since,

$$\begin{aligned}\frac{d}{dt}\langle\Omega\rangle &= \frac{d}{dt}\int d\vec{x}\psi^*(\vec{x})\Omega\psi(\vec{x}) \\ &= \int d\vec{x}\left(\frac{\partial\psi^*}{\partial t}\Omega\psi + \psi^*\frac{\partial\Omega}{\partial t}\psi + \psi^*\Omega\frac{\partial\psi}{\partial t}\right),\end{aligned} \tag{2.15}$$

and

$$\frac{\partial\psi}{\partial t} = -\frac{i}{\hbar}H\psi, \tag{2.16}$$

it follows that the time derivative $\dot{\Omega}$ of an operator Ω is given by

$$\dot{\Omega} = \frac{\partial\Omega}{\partial t} + \frac{i}{\hbar}[H, \Omega]. \tag{2.17}$$

This is known as Heisenberg's equation of motion.

9. States of a closed system are described by wavefunctions. However, this is not true of systems that are parts of a larger system and which can be called a subsystem. If we have an operator Ω which operates on the coordinate x of the subsystem and not on the remaining coordinates q of the total system, then

$$\langle \Omega \rangle = \int dx \int dq \psi^*(q,x)\Omega\psi(q,x) = \int dx \left(\Omega\rho(x,x')\right)_{x=x'}, \tag{2.18}$$

where

$$\rho(x,x') = \rho^*(x',x) = \int dq \psi^*(q,x)\psi(q,x') \tag{2.19}$$

is the density matrix. Thus, a subsystem is described by the density matrix.

2.2 Uncertainty Relations

Heisenberg's uncertainty relation discussed in Chapter 1 can be obtained from the ground rules presented in the previous section. For this, we consider a particle of momentum $\vec{p}_0$ confined to a small volume, whose wavefunction

$$\psi(\vec{p}_0,\vec{r}) = u(\vec{r})e^{\frac{i}{\hbar}\vec{p}_0\cdot\vec{r}}, \tag{2.20}$$

where $u(\vec{r})$ differs from zero within this volume. The probability amplitude of momentum of the particle is given by the Fourier transform

$$\phi(\vec{p}) = \int d^3re^{-\frac{i}{\hbar}\vec{p}\cdot\vec{r}}\psi(\vec{p}_0,\vec{r}) = \int d^3re^{\frac{i}{\hbar}(\vec{p}_0-\vec{p})\cdot\vec{r}}u(\vec{r}). \tag{2.21}$$

The contribution to the integral will essentially come from momenta $|\Delta\vec{p}| = |(\vec{p}-\vec{p}_0)|$, which is small compared to $\hbar/|\Delta\vec{r}|$ because for a large period of the oscillatory part $e^{\frac{i}{\hbar}(\vec{p}_0-\vec{p})\cdot.\vec{r}}$, the integrand will contribute almost nothing. Therefore we will have

$$\Delta\vec{p}\cdot\Delta\vec{x} \geq \hbar. \tag{2.22}$$

A more precise relation is obtained from the inequality

$$\int_{-\infty}^{\infty} dx \left|\left(\alpha x\psi + \frac{d\psi}{dx}\right)\right|^2 \geq 0, \tag{2.23}$$

where α is an arbitrary real constant. In terms of root mean square (standard) deviations Δx and Δp_x,

$$\begin{aligned}(\Delta x)^2 &= \langle (x - \langle x \rangle)^2 \rangle = \langle x^2 \rangle,\\ (\Delta p_x)^2 &= \langle (p_x - \langle p_x \rangle)^2 \rangle = \langle p^2 \rangle,\\ \langle x \rangle &= 0, \langle p_x \rangle = 0.\end{aligned} \tag{2.24}$$

The inequality (2.23) can be written as,

$$\alpha^2(\Delta x)^2 - \alpha + \frac{1}{\hbar^2}(\Delta p)^2 \geq 0. \tag{2.25}$$

It is satisfied if

$$\Delta x \Delta p_x \geq \frac{h}{2}. \tag{2.26}$$

The minimum uncertainty,

$$\Delta x \Delta p_x = \frac{h}{2}, \tag{2.27}$$

occurs for the state with wavefunction

$$\psi(x) = \frac{1}{(2\pi)^{1/4}(\Delta x)^{1/2}} \exp\left[-\frac{x^2}{4(\Delta x)^2}\right]. \tag{2.28}$$

It is on account of the uncertainty in simultaneous measurement of the position and momentum that quantum mechanics cannot give as complete description of a physical system as classical mechanics does, where such a description of the system can be made if the coordinates and momenta of its constituents are known at a given moment of time, with the equations of motion determining its subsequent behaviour. This is not possible in quantum mechanics since a particle does not have a definite path on account of the uncertainty principle.

2.3 Hidden Variable Theory

As an alternative to quantum mechanics, David Bohm proposed a theory where hidden variables are introduced in a classical deterministic theory. The lack of simultaneous, well-defined position and momentum is due to hidden variables over which classical average has to be taken. An apt analogy, as given by Bohm, is thermodynamics, where we have deterministic equations of state involving pressure, volume and temperature of the system. However, near a critical point these quantities no longer obey an equation of state—the deterministic laws of thermodynamics break down. One has to deal with the system in terms of the position and momentum of individual atoms or molecules which can be considered as hidden variables in thermodynamics, over which averages have to be taken to obtain the physical laws. These variables cannot be observed by thermodynamics methods alone.

A hidden variable theory in which there will be no uncertainty will be deterministic. So far, no one knows what these variables are. However, such theories will not give the results of quantum mechanics. John von Neuman has proved a theorem to this effect. However, this proof is believed not to be fool-proof as shown by Bell's analysis. The situation is much more subtle.

2.4 Schrödinger, Heisenberg and Dirac Pictures

We have seen in section 1 of this chapter that the time-dependence of the wavefunction is determined by the Schrödinger equation:

$$i\hbar\frac{\partial\psi}{\partial t} = H\psi, \tag{2.29}$$

and those of operators given by the Heisenberg equation:

$$\dot{\Omega} = \frac{\partial\Omega}{\partial t} + \frac{i}{\hbar}[H, \Omega]. \tag{2.30}$$

Actually these two equations refer to two different pictures of quantum mechanics and should not be used simultaneously. In the former the operators are time-independent and in the latter the wavefunctions are time-independent. The two pictures are known as Schrödinger and Heisenberg pictures respectively. The operators and wavefunctions in the two pictures are related by,

$$\psi_s(t) = e^{-\frac{iHt}{\hbar}}\psi_H(t), \quad \frac{\partial\psi_H}{\partial t} = 0,$$

$$\Omega_s(t) = e^{-\frac{iHt}{\hbar}}\Omega_H(0)e^{\frac{iHt}{\hbar}}, \quad \frac{d\Omega_s}{dt} = 0. \tag{2.31}$$

A picture intermediate between these two is used extensively in perturbation theory, according to which

$$\begin{aligned} i\hbar\frac{\partial\psi_D(t)}{\partial t} &= H_I\psi_D(t) \\ \frac{d\Omega_D}{dt} &= \frac{\partial\Omega_D}{\partial t} + \frac{i}{\hbar}[H_0, \Omega_D], \end{aligned} \tag{2.32}$$

where the Hamiltonian H consists of two parts:

$$H = H_0 + H_I. \tag{2.33}$$

H_0 describes the unperturbed part and H_I represents the interaction. This picture is commonly known as the interaction picture and sometimes as the Dirac picture. In this picture, unlike the Schrödinger and Heisenberg pictures, both wavefunction and operators are time-dependent. The relation between the Dirac picture wavefunction and operators and the Schrödinger picture is:

$$\psi_D(t) = e^{+\frac{i}{\hbar}H_0 t}\psi_s(t), \Omega_D(t) = e^{+\frac{i}{\hbar}H_0 t}\Omega_s e^{-\frac{i}{\hbar}H_0}. \tag{2.34}$$

Problems

1. Decide which of the following operators are hermitian using the definition in the text.
(a) $ix\frac{d}{dx}$ (b) $\left[x^2\left(\frac{d}{dx}\right)^2 + \left(\frac{d}{dx}\right)^2 x^2\right]$ (c) $\left(\frac{d}{dx}\right)^2$ (d) $\left(\frac{d}{dx}\right)^3$

2. Find the linear combination of the three wavefunctions $\psi_1 = \sin\theta e^{i\phi}$, $\psi_2 = \sin\theta e^{-i\phi}$, $\psi_3 = \frac{1}{\sqrt{2}}\cos\theta$, for which

$$L_y = \frac{1}{\sqrt{2}}\begin{pmatrix} 0 & -i & 0 \\ i & 0 & -i \\ 0 & i & 0 \end{pmatrix}$$

has an eigenvalue $\lambda = -1$.
Hint: The coefficients of the linear combination can be obtained by finding the eigenvectors of the equation $L_y\psi = -\psi$.

3. If operators A and B are hermitian, under what conditions will the product operator AB be hermitian?

4. For $\psi(x) = e^{-\alpha^2 x^2}$ and $\psi(x) = \frac{1}{(2\pi)^{1/4}}\frac{1}{(\Delta x)^{1/2}}\exp[-x^2/4(\Delta x)^2]$, calculate $\Delta x\Delta p$.

Bibliography

[1] N. Bohr, *Naturwiss*, **16**, 245 (1929); **17**, 483 (1929); **18**, 73 (1930).
[2] W. Heisenberg, *Z. Physik*, **3**, 1977 (1927).
[3] W. Heisenberg, *The Physical Principles of Quantum Theory*, University of Chicago Press (1930).
[4] Albert Einstein, Philosopher Scientist, ed. P.A. Schipp, (Tudor Publishing Company, New York 1949 and 1951). Articles by N. Bohr and A. Einstein.
[5] D. Bohm, *Quantum Theory*, Prentice Hall Inc., New Jersey (1951).

3 Symmetry and Conservation Laws

3.1 Introduction

Conservation of dynamical variables are associated with symmetries of the Hamiltonian. If it is invariant under a symmetry operation, then the generator is a constant of motion. Symmetry transformation has to be unitary for the expectation values of the quantum mechanical operator undergoing such transformation to remain unchanged. For this to be satisfied, under such a transformation, the wavefunction ψ and the operator Ω have to transform as

$$\psi \longrightarrow \psi' = U\psi, UU^\dagger = U^\dagger U = 1 \tag{3.1}$$

$$\Omega \longrightarrow \Omega' = U\Omega U^\dagger. \tag{3.2}$$

In that case

$$\langle\Omega\rangle = \langle\psi|\Omega|\psi\rangle,$$

$$\langle\Omega'\rangle = \langle\psi'|\Omega'|\psi'\rangle = \langle\psi|U^\dagger U\Omega U^\dagger U|\psi\rangle \tag{3.3}$$

and

$$\langle\Omega\rangle = \langle\Omega'\rangle \quad \text{if} \quad U^\dagger U = 1, \tag{3.4}$$

i.e., the transformation is unitary. Since U is unitary we can express it in terms of a hermitian operator $\vec{F}$:

$$U = e^{i\vec{\alpha}\cdot\vec{F}}, \vec{F} = \vec{F}^\dagger \tag{3.5}$$

where $\vec{\alpha}$ is a parameter of the transformation. The hermitian operator $\vec{F}$ is called the generator of the transformation. For infinitesimal transformations, $\vec{\alpha}$ is small so that

$$U = 1 + i\vec{\alpha}\cdot\vec{F}. \tag{3.6}$$

Under such a transformation, an operator Ω changes by

$$\begin{aligned}\delta\Omega &= (\Omega' - \Omega) = (1 + i\alpha_i F_j)\Omega(1 - i\alpha_i F_j) - \Omega \\ &= i\alpha_i[F_i, \Omega]\end{aligned} \tag{3.7}$$

to first order in $\vec{\alpha}$. For the Hamiltonian operator,

$$\delta H = i\alpha_i[F_i, H]. \tag{3.8}$$

It is thus clear that if the Hamiltonian is invariant under the transformation, i.e., $\delta H = 0$, then

$$[F_i, H] = i\hbar\frac{dF_i}{dt} = 0 \tag{3.9}$$

if $\frac{\partial F_i}{\partial t} = 0$, i.e., F_i do not contain time explicitly. Thus the generators F_i are time independent. In other words they are conserved.

3.2 Conservation of Energy

It will be shown that energy is conserved if the Hamiltonian is invariant under infinitesimal time translation:

$$t \longrightarrow t' = t + \delta t, \tag{3.10}$$

where $\delta t = \alpha$.
Under this operation,

$$\psi(t) \longrightarrow \psi(t + \delta t) = \psi(t) + \delta t\frac{\partial\psi}{\partial t}, \tag{3.11}$$

$$\Omega(t) \longrightarrow \Omega(t + \delta t) = \Omega(t) + \delta t\frac{\partial\Omega}{\partial t}. \tag{3.12}$$

Comparison of (3.11) with (3.1) gives

$$F = -\frac{1}{\hbar}H \tag{3.13}$$

on account of the equation $\frac{\partial\psi}{\partial t} = -\frac{i}{\hbar}H\psi$.
Thus the Hamiltonian is the generator of infinitesimal time translation, and since it commutes with itself, it is time-independent, i.e., energy is conserved. For finite-time translation,

$$\psi(t + \tau) \longrightarrow e^{\frac{i\tau H}{\hbar}}\psi(t). \tag{3.14}$$

3.3 Conservation of Momentum

By a procedure similar to the previous section, we shall show that momentum is conserved if the Hamiltonian is invariant under infinitesimal space translation defined by

$$\vec{r} \longrightarrow= \vec{r} + \delta\vec{r}, \delta\vec{r} = \vec{\alpha}, \tag{3.15}$$

under which

$$\psi(\vec{r}) \longrightarrow \psi(\vec{r} + \delta\vec{r}) = \psi(\vec{r}) + \delta\vec{r} \cdot \vec{\nabla}\psi(\vec{r}) \tag{3.16}$$

$$\Omega(\vec{r}) \longrightarrow \Omega(\vec{r} + \delta\vec{r}) = \Omega(\vec{r}) + \delta\vec{r} \cdot \vec{\nabla}\Omega(\vec{r}). \tag{3.17}$$

Comparison with (3.7) gives

$$\vec{F} = -i\vec{\nabla} = \frac{\vec{p}}{\hbar}, \tag{3.18}$$

which is conserved if H is invariant under transformation (3.15). For finite translation

$$\psi(\vec{r} + \vec{R}) = e^{\frac{i\vec{p}\cdot\vec{R}}{\hbar}}\psi(\vec{r}). \tag{3.19}$$

3.4 Conservation of Angular Momentum

The conservation of angular momentum is a consequence of the invariance of Hamiltonian under an infinitesimal rotation

$$\vec{r} \longrightarrow \vec{r} - \vec{\theta} \times \vec{r}, \tag{3.20}$$

where $|\vec{\theta}|$ is an infinitesimally small angle. Under this rotation,

$$\psi(\vec{r}) \longrightarrow \psi(\vec{r} - \vec{\theta} \times \vec{r}) = \psi(\vec{r}) - (\vec{\theta} \times \vec{r}) \cdot \vec{\nabla}\psi(\vec{r}) \tag{3.21}$$

so that

$$\vec{F} = -i\vec{r} \times \vec{\nabla} = \frac{\vec{r} \times \vec{p}}{\hbar} = \frac{\vec{L}}{\hbar}, \tag{3.22}$$

which shows that angular momentum is conserved if H is invariant under an infinitesimal rotation. For a finite rotation

$$\vec{r}' = e^{\left(i\frac{\theta L}{\hbar}\right)}\vec{r},$$

$$\psi(\vec{r}') = e^{i\frac{\vec{\theta}\vec{L}}{\hbar}}\psi(\vec{r}). \tag{3.23}$$

3.5 Conservation of Parity

The transformation associated with conservation of parity is a discrete transformation

$$\vec{r} \longrightarrow -\vec{r} \tag{3.24}$$

as opposed to the previously considered continuous transformations. No spatial rotation can cause such a transformation. Under this,

$$\psi(\vec{r}) \longrightarrow \psi(-\vec{r}) = P\psi(\vec{r}) \tag{3.25}$$

and hence

$$P^2\psi(\vec{r}) = P\psi(-\vec{r}) = \psi(\vec{r}) \tag{3.26}$$

thereby giving, for an eigenfunction,

$$P\psi(\vec{r}) = \pm\psi(\vec{r}). \tag{3.27}$$

Thus the eigenvalues of the space inversion operator are ± 1 and will be conserved if the Hamiltonian commutes with P, i.e., H is invariant under this inversion. This means that if a closed system has a state with a definite parity, it will not change with time.

3.6 Conservation of Probability

It has been noted in the previous chapter that the conservation of probability follows from the Schrödinger equation. In this section we shall show that there is an underlying symmetry behind this conservation. This symmetry is the invariance of Schrödinger equation under the transformation

$$\psi(\vec{r}) \longrightarrow \psi'(\vec{r}) = e^{\frac{i\alpha}{\hbar}(\vec{r})}\psi(\vec{r}). \tag{3.28}$$

Form-invariance requires that the Schrödinger equation for $\psi'(\vec{r})$ be

$$i\hbar\frac{\partial\psi'(\vec{r})}{\partial t} = -\frac{\hbar^2}{2m}\nabla^2\psi'(\vec{r}) + V(r)\psi'(\vec{r}). \tag{3.29}$$

Substitution of (3.28) in this equation gives

$$\begin{aligned} &\frac{\partial\alpha}{\partial t}\psi - i\hbar\frac{\partial\psi}{\partial t} + \frac{1}{2m}(\vec{\nabla}\alpha)^2\psi - \frac{i\hbar}{2m}\nabla^2\alpha\psi \\ &\quad - \frac{i\hbar}{m}(\vec{\nabla}\alpha)\cdot(\vec{\nabla}\psi) - \frac{\hbar^2}{2m}\nabla^2\psi + V\psi = 0, \end{aligned} \tag{3.30}$$

which on account of the Schrödinger equation

$$i\hbar\frac{\partial\psi(\vec{r})}{\partial t} = -\frac{\hbar^2}{2m}\psi(\vec{r}) + V(r)\psi(\vec{r}) \tag{3.31}$$

becomes

$$\frac{\partial\alpha}{\partial t}\psi + \frac{1}{2m}(\vec{\nabla}\alpha)^2\psi - \frac{i\hbar}{2m}\nabla^2\alpha\psi - \frac{i\hbar}{m}(\vec{\nabla}\alpha)\cdot(\vec{\nabla}\psi) = 0. \tag{3.32}$$

Multiplication of this equation on the left by $\psi^*(\vec{r})$ gives

$$\frac{\partial\alpha}{\partial t}\rho + \frac{1}{2m}(\vec{\nabla}\alpha)^2\rho - \frac{i\hbar}{2m}(\nabla^2\alpha)\rho - \frac{i\hbar}{m}(\psi^*\vec{\nabla}\psi)\cdot\vec{\nabla}\alpha = 0, \tag{3.33}$$

where $\rho = \psi^*\psi$, the probability density is real. Equating real and imaginary terms of this equation separately to zero gives,

$$(\nabla^2\alpha)\rho + 2(\vec{\nabla}\cdot\alpha)(\psi^*\vec{\nabla}\psi) = 0 \tag{3.34}$$

$$\left(\frac{\partial\alpha}{\partial t}\right) + \frac{1}{2m}(\vec{\nabla}\alpha)^2 = 0. \tag{3.35}$$

Equation (3.35) is the Hamilton–Jacobi equation if we identify α with action S. We now take the complex conjugate of (3.34) which is

$$(\nabla^2\alpha)\rho + 2(\vec{\nabla}\alpha)\cdot(\psi\vec{\nabla}\psi^*) = 0. \tag{3.36}$$

On adding (3.34) and (3.36) we get

$$(\nabla^2\alpha)\rho + (\vec{\nabla}\cdot\vec{\alpha})[\psi^*(\vec{\nabla}\psi) + (\vec{\nabla}\psi^*)\psi] = 0, \tag{3.37}$$

which can be written as

$$\vec{\nabla}\cdot\left(\rho\frac{\vec{\nabla}\alpha}{m}\right) = \vec{\nabla}\left(\rho\frac{\vec{p}}{m}\right) = \vec{\nabla}\cdot(\rho\vec{v}) = (\vec{\nabla}\cdot\vec{j}) = 0. \tag{3.38}$$

The continuity equation

$$\frac{\partial\rho}{\partial t} + \vec{\nabla}\cdot\vec{j} = 0, \tag{3.39}$$

which follows from Schrödinger equation, then leads to conservation of probability:

$$\frac{\partial P}{\partial t} = \int d^3r\frac{\partial\rho}{\partial t} = 0. \tag{3.40}$$

Normally, conservation follows from equation of continuity by taking its volume integral, converting the volume integral over divergence of the current to a surface integral and then demanding that the current vanishes at infinity. Here no such condition is necessary.

Problems

1. Show that the Schrödinger equation

$$i\hbar\frac{\partial\psi}{\partial t}+\frac{1}{2m}\nabla^2\psi+V\psi=0$$

is invariant under the Galilean transformation

$$\vec{x}\longrightarrow\vec{x}-\vec{v}t$$

$$t\longrightarrow t$$

but not so under the Lorentz transformation

$$\vec{x}\longrightarrow\frac{\vec{x}-\vec{v}t}{\sqrt{1-\frac{v^2}{c^2}}},$$

$$t\longrightarrow\frac{t-\vec{v}\cdot\vec{x}/c^2}{\sqrt{1-\frac{v^2}{c^2}}}.$$

What is conserved as a consequence of the Galilean invariance?

2. Show that the Schrödinger equation with potential energy $= gV(r)$ and the commutation relations $[q_i, p_j] = i\hbar\delta_{ij}$ are invariant under the transformations:

$$\vec{r}\longrightarrow\lambda\vec{r}$$

$$t\longrightarrow\lambda^2 t$$

$$g\longrightarrow g(\lambda)^{-(d+2)}$$

$$V(\vec{r})\longrightarrow\lambda^d V(\vec{r})$$

3. Show that the generator of the transformation in the previous problem is

$$F=\frac{i}{\hbar}\left[tdH+\vec{r}\cdot\vec{p}-\frac{d+2}{2}\int dt\ mv^2\right].$$

4. Show that the radial Schrödinger equation

$$\frac{d^2R}{dr^2}+\frac{2}{r}\frac{dR}{dr}-\left[\frac{l(l+1)+\beta}{r^2}\right]R=0$$

is invariant under the transformation

$$\vec{r} \longrightarrow \lambda \vec{r} \; .$$

Obtain the differential form of its generator and its eigenvalue and eigenfunction.

Bibliography

[1] L.D. Landau and E.M. Lifshitz, *Quantum Mechanics*, Third Edition, Pergamon Press, Oxford (1977).

[2] D. Bohm, *Quantum Theory*, Prentice Hall Inc., New Jersey (1951).

[3] L.I. Schiff, *Quantum Mechanics*, 3rd edition McGraw Hill Book Company Inc., New York (1977).

4 Energy, Momentum and Angular Momentum

4.1 Energy

We have seen in chapter 3 that the Hamiltonian is the generator of time translation and if the Hamiltonian of a system is invariant under this transformation, its energy is conserved. What this means is that for such systems, all times are equivalent, i.e., the energy is the same for all times. Such states, for which energy has a definite value at all times, are called stationary states and for them

$$H|\psi_n\rangle = E_n|\psi_n\rangle \tag{4.1}$$

where E_n are energy eigenvalues and the equation

$$i\hbar\frac{\partial}{\partial t}|\psi_n(\vec{r},)\rangle = H|\psi_n(\vec{r},t)\rangle \tag{4.2}$$

has the solution

$$|\psi_n(\vec{r},t)\rangle = e^{-\frac{iE_nt}{\hbar}}|u_n(\vec{r})\rangle. \tag{4.3}$$

For an arbitrary wavefunction which is a linear superposition of energy eigenfunctions, we have, by Taylor's expansion followed by application of (4.2),

$$\begin{aligned}\psi(\vec{r},t) &= \psi(\vec{r},0) + t\frac{\partial\psi(\vec{r},t)}{\partial t}\bigg|_{t=0} + \frac{t^2}{2!}\frac{\partial^2\psi(\vec{r},t)}{\partial t^2}\bigg|_{t=0} + \cdots \\ &= \psi(\vec{r},0) + \frac{it}{\hbar}H\psi(\vec{r},0) - \frac{t^2}{2!\hbar^2}H^2\psi(\vec{r},0) + \cdots \\ &= e^{-\frac{iHt}{\hbar}}\psi(\vec{r},0).\end{aligned} \tag{4.4}$$

If an operator commutes with the Hamiltonian, its expectation value does not change with time. For such operators we must have, on account of (4.4),

$$O(\vec{r},t) = e^{-\frac{iHt}{\hbar}}O(\vec{r},0)e^{\frac{i}{\hbar}Ht}. \tag{4.5}$$

The energy eigenvalues of a system can be discrete, continuous or a combination of both. For discrete states, $\int d^3r|\psi(\vec{r})|^2$ is finite. For this to happen $|\psi(\vec{r})|^2$ must decrease with increasing $|\vec{r}|$, which implies that the probability of finding the system in the state with wavefunction ψ at infinity vanishes. Such states are called bound states. For states with continuous energy eigenvalues, $\int d^3r|\psi(\vec{r})|$ diverges and there is a finite probability of finding the system in such states at infinity.

4.2 Momentum

Momentum is the generator of space translation and is conserved if the Hamiltonian of the system is invariant under such a translation. It is represented by the differential operator $\vec{\nabla}$ (apart from a constant)

$$\vec{p}_{\text{op}} = -i\hbar\vec{\nabla}. \tag{4.6}$$

The eigenvalue equation

$$-i\hbar\vec{\nabla}\psi = \vec{p}\psi \tag{4.7}$$

has the solution

$$\psi_p(\vec{r}) = \frac{1}{(2\pi)^{3/2}} e^{\frac{i\vec{p}\cdot\vec{r}}{\hbar}}, \tag{4.8}$$

where the eigenvalue of $\vec{p}$ has continuous values. The normalization for the corresponding eigenfunctions is expressed in terms of Dirac delta function

$$\int d^3r\psi^*_{p'}(\vec{r})\psi_p(\vec{r}) = \delta^3\left(\frac{\vec{p}-p'}{\hbar}\right). \tag{4.9}$$

As in the previous section, one can have, for a linear superposition $\psi(\vec{r})$ of momentum eigenfunctions,

$$\psi(\vec{r}) = e^{i\vec{p}_{\text{op}}\cdot(\vec{r}-\vec{r}')}\psi(\vec{r}') \tag{4.10}$$

and

$$O(\vec{r}) = e^{i\vec{p}_{\text{op}}(\vec{r}-\vec{r}')}O(\vec{r}')e^{-i\vec{p}_{\text{op}}(\vec{r}-\vec{r}')}. \tag{4.11}$$

4.3 Angular Momentum

De nition and commutation relations

Angular momentum is the generator of space rotations and is conserved if the Hamiltonian is invariant under such rotations. It is defined as

$$\vec{L} = \vec{r}\times\vec{p} = -i\hbar(\vec{r}\times\vec{\nabla}). \tag{4.12}$$

Since $\vec{r}$ and $\vec{p}$ do not commute, the components of $\vec{L}$ do not commute. On using the commutation relation

$$[r_i, p_j] = i\hbar\delta_{ij}, \tag{4.13}$$

we find

$$\begin{aligned} [L_i, r_j] &= i\mathcal{E}_{ijk} r_k \\ [L_i, p_j] &= i\mathcal{E}_{ijk} p_k \\ [L_i, L_j] &= i\mathcal{E}_{ijk} L_k, \end{aligned} \tag{4.14}$$

where $\mathcal{E}_{ijk}$ is the totally antisymmetric unit tensor of rank three. Further, we find that $L^2 = L_1^2 + L_2^2 + L_3^2$ commutes with all the three components, L_1, L_2, L_3:

$$[L^2, L_i] = 0 \ , \qquad\qquad i = 1, 2, 3. \tag{4.15}$$

It is often found convenient to make use of the operators

$$L_\pm = L_1 \pm iL_2. \tag{4.16}$$

Their commutation relations with themselves and with L_3 work out to be

$$\begin{aligned} [L_+, L_-] &= 2L_3 \\ [L_3, L_+] &= L_+ \\ [L_3, L_-] &= -L_-. \end{aligned} \tag{4.17}$$

Since the components of $\vec{L}$ do not commute among themselves, they cannot be simultaneously determined. However, since L^2 and L_3 commute, they can have common eigenfunctions.

Matrix elements

In spherical polar coordinates

$$L_3 = -i\frac{\partial}{\partial\varphi} \ , \hbar = 1. \tag{4.18}$$

The eigenvalue equation for this operator is

$$-i\frac{\partial\psi}{\partial\varphi} = m\psi, \tag{4.19}$$

whose solution is

$$\psi = Ne^{im\varphi}, \tag{4.20}$$

where N is a constant normalisation factor. Since this eigenfunction has to be single-valued,

$$\psi_m(\varphi) = \psi_m(\varphi + 2\pi). \tag{4.21}$$

We obtain

$$m = 0\,, \pm 1,\ \ \pm\, 2\,, \ldots . \tag{4.22}$$

The quantum number m is called the magnetic quantum number. The normalisation factor N is easily seen to be

$$N^2 = \frac{1}{2\pi}. \tag{4.23}$$

Since L_1 and L_2 do not commute with each other but commute with the Hamiltonian, as per discussion in chapter 2, the states with different m will have the same energy leading to what we may call the m-degeneracy.

To obtain the eigenvalue of the square of the angular momentum operator L^2, we note that since

$$[L^2, L_3] = 0, \tag{4.24}$$

if ψ is an eigenfunction of L^2 with eigenvalue λ_l,

$$L^2\psi = \lambda_l \psi; \tag{4.25}$$

then

$$L^2(L_3\psi) = L_3 L^2 \psi = \lambda_l(L_3\psi), \tag{4.26}$$

which shows that $L_3\psi$ is also an eigenfunction of L^2 with the same eigenvalue. Further, we note that on account of commutation relation (4.17)

$$[L_+ L_3] = -L_+,$$

the eigenvalue equation

$$L_+(L_3\psi_m) - m(L_+\psi_m) \tag{4.27}$$

leads to

$$L_3(L_+\psi_m) = (m+1)(L_+\psi_m), \tag{4.28}$$

which tells us that $L_+\psi_m$ is an eigenfunction of L_3 with eigenvalue $(m+1)$. Thus the operator L_+ raises the value of m by one unit. With exactly the same type of argument, one gets

$$L_3(L_-\psi_m) = (m-1)(L_-\psi_m), \tag{4.29}$$

which tells us that the operator L_- lowers the value of m by one unit. One can, by repeated application of L_+, get as large a positive value of m as possible and by repeated application of L_-, get as large a negative value of m as possible. On the other hand, from

$$L^2 = L_1^2 + L_2^2 + L_3^2 = L_1^2 + L_2^2 + m^2, \tag{4.30}$$

it is seen that

$$m^2 \leq L^2, \tag{4.31}$$

which shows that $|m|$ cannot grow beyond $\sqrt{L^2}$. We therefore impose the conditions

$$L_+\psi_{m_1} = 0, \quad L_-\psi_{m_2} = 0, \tag{4.32}$$

where m_1 is the maximum positive value and m_2 the maximum negative value of L_3 for a given value of L^2. It then follows that

$$\begin{aligned} L^2\psi_{m_1} &= m_1(m_1+1)\psi_{m_1}, \\ L^2\psi_{m_2} &= m_2(m_2-1)\psi_{m_2}. \end{aligned} \tag{4.33}$$

For both these equations to be true we must have

$$m_2 = -m_1. \tag{4.34}$$

Designating m_1 by the integer l so that m takes values $-l, -l+1, \ldots l$, we get

$$L^2\psi_l = l(l+1)\psi_l. \tag{4.35}$$

The quantum number $(l+1)$ is called azimuthal quantum number. The eigenfunctions of L^2 are obtained by making use of the expression

$$L^2 = -\left[\frac{1}{\sin^2\theta}\frac{\partial^2}{\partial\varphi^2} + \frac{1}{\sin\theta}\frac{\partial}{\partial\theta}\left(\sin\theta\frac{\partial}{\partial\theta}\right)\right] \tag{4.36}$$

in spherical polar coordinates. Its eigenfunctions are characterised by quantum numbers l and m:

$$\psi_{lm}(\theta\varphi) = Y_{l,m}(\theta,\varphi) = \Theta_{l,m}(\theta)\Phi_m(\varphi), \tag{4.37}$$

where $Y_{l,m}(\theta,\varphi)$ are the spherical harmonics. The eigenvalue equation

$$L^2\psi_{lm} = l(l+1)\psi_{lm} \tag{4.38}$$

then becomes

$$\frac{1}{\sin\theta}\frac{d}{d\theta}\left(\sin\theta\frac{d\Theta_{lm}(\theta)}{d\theta}\right) - \frac{m^2}{\sin^2\theta}\Theta_{lm}(\theta) + l(l+1)\Theta_{lm} = 0 \tag{4.39}$$

whose normalised solutions are

$$\Theta_{lm}(\theta) = (-1)^m\sqrt{\frac{(2l+1)(l-|m|)!}{2(l+|m|)!}}\; P_l^{|m|}(\cos\theta), \tag{4.40}$$

where $P_l^m(\cos\theta)$ are the associated Legendre polynomials.

We now evaluate the matrix elements of L_+ and L_-. For this, we take the diagonal matrix elements of the identity

$$L^2 = L_+L_- + L_3^2 - L_3, \tag{4.41}$$

which gives

$$\begin{aligned} l(l+1) &= \langle m|L_+|m-1\rangle\langle m-1|L_-|m\rangle + m^2 - m \\ &= |\langle m|L_+|m-1\rangle|^2 + m(m-1). \end{aligned} \tag{4.42}$$

Use has been made of the identity

$$\langle m-1|L_-|m\rangle = \langle m|L_+|m-1\rangle^* \tag{4.43}$$

obtained from hermiticity of the operators L_1 and L_2. From (4.42) we get

$$\langle m-1|L_-|m\rangle = \langle m|L_+|m-1\rangle^* = \sqrt{(l+m)(l-m+1)}. \tag{4.44}$$

Addition of angular momenta

Consider a system with angular momentum $\vec{L}$, consisting of two subsystems with angular momenta $\vec{L}(1)$ and $\vec{L}(2)$ such that

$$\vec{L} = \vec{L}(1) + \vec{L}(2), \tag{4.45}$$

$$[L_i(1), L_j(2)] = 0, \tag{4.46}$$

with

$$\begin{aligned} &\vec{L}^2(1)|l_1m_1|\rangle = l_1(l_1+1)|l_1m_1\rangle\;,\; \vec{L}^2(2)|l_2m_2\rangle = l_2(l_2+1)|l_2m_2\rangle \\ &L_3(1)|l_1m_1\rangle = m_1|l_1m_1\rangle\;,\; L_3(2)|l_2m_2\rangle = m_2|l_2m_2\rangle \\ &L^2|lm\rangle = l(l+1)|lm\rangle\;,\; L_3|lm\rangle = m|lm\rangle. \end{aligned} \tag{4.47}$$

An example of such a system is an atom with two electrons. There are two sets of commuting operators:

$$\vec{L}^2(1),\;\; L_3(1),\;\; \vec{L}^2(2),\;\; L_3(2)$$

and

$$\vec{L}^2(1),\ \vec{L}^2(2),\ \vec{L}^2,\ L_3, \tag{4.48}$$

the two having common eigenstates $|l_1 l_2 m_1 m_2\rangle$ and $|l_1 l_2 l m\rangle$ respectively, where

$$|l_1 l_2 m_1 m_2\rangle = |l_1 m_1\rangle |l_2 m_2\rangle. \tag{4.49}$$

These two eigenstates are related by a unitary transformation

$$|l_1 l_2 l m\rangle = \sum_{m_1 m_2} \langle l_1 l_2 m_1 m_2 | l_1 l_2 l m\rangle |l_1 l_2 m_1 m_2\rangle \tag{4.50}$$

with the constraint $m_1 + m_2 = m$ on account of (4.45) and (4.47). The coefficients $\langle l_1 l_2 m_1 m_2 | l_1 l_2 l m\rangle$ are called Clebsch–Gordan (CG) coefficients. The phases of the states are so chosen that these coefficients are real, i.e.,

$$\langle m_1 m_2 | l m\rangle = \langle l m | m_1 m_2\rangle. \tag{4.51}$$

The indices $l_1 l_2$ occurring in the states have been dropped. The inverse of the transformation (4.50) is

$$|m_1 m_2\rangle = \sum_{l=|l_1-l_2|}^{l_1+l_2} \langle l m | m_1 m_2\rangle |l m\rangle. \tag{4.52}$$

The CG coefficients can be expressed in terms of $3j$ symbol

$$\langle m_1 m_2 | l m\rangle = (-)^{l_1 - l_2 + m} \sqrt{2l+1} \begin{pmatrix} l_1 & l_2 & l \\ m_1 & m_2 & -m \end{pmatrix}. \tag{4.53}$$

Since the transformation (4.50) is unitary, the CG coefficients obey orthogonality relation

$$\sum_{m_1 m_2} \langle m_1 m_2 | l m\rangle \langle m_1 m_2 | l' m'\rangle = \delta_{ll'} \delta_{mm'}, \tag{4.54}$$

$$\sum_{l} \langle m_1 m_2 | l m\rangle \langle m_1' m_2' | l m\rangle = \delta_{m_1 m_1'} \delta_{m_2 m_2'}. \tag{4.55}$$

The general form of $3j$ symbols was first calculated by Wigner and later expressed in symmetrical from by Racah. The general expression is lengthy. From this expression one gets for a particular CG coefficient

$$\begin{aligned} &\langle m_1 m_2 | l_1 + l_2, m\rangle \\ &\quad = \left[\frac{(2l)!(2l_2)!(l_1+l_2+m)!(l_1+l_2-m)!}{(2l_1+2l_2+1)!(l_1+m)!(l_1-m_1)!(l_2+m_2)!(l_2-m_2)!} \right]^{\frac{1}{2}} \\ &\qquad \times (-1)^{l_1 - l_2 + m}. \end{aligned} \tag{4.56}$$

Matrix elements of tensor operators

An irreducible tensor T_{LM} of rank L has $(2L + 1)$ components which transform like the components of spherical harmonics Y_{LM}.

Its matrix elements are defined by

$$T_{LM}\psi_{lm} = \sum_{l'm'} \langle l'm'|T_{LM}|lm\rangle \psi_{l'm'}. \tag{4.57}$$

Since both sides of this equation have the same transformation properties as those of (4.52), it follows that all matrix elements of T_{LM} are zero except for

$$m' = m + M. \tag{4.58}$$

Hence

$$T_{LM}\psi_{lm} = \alpha \sum_{l',m'} \langle l', (M + m)|Mm\rangle \psi_{l',M+m}, \tag{4.59}$$

where α is a constant parameter and which on comparing with (4.57) gives

$$\langle l', (m + M)|T_{LM}|lm\rangle = \alpha\langle l', (m + M)|Mm\rangle. \tag{4.60}$$

Expressed in terms of $3j$ symbol as per (4.53), this relation becomes

$$\langle l'm'|T_{LM}|lm\rangle = (-1)^{l'-L+m}\sqrt{2l+1}\begin{pmatrix} l' & L & l \\ M+m & M & -m \end{pmatrix}\langle l'||T_L||l\rangle, \tag{4.61}$$

where $\left\langle l'||T_L||l\right\rangle = \alpha$ is called the reduced matrix element and depends on the dynamics of the system. Relation (4.61) is the Wigner–Eckart theorem.

Matrix elements of vector operators $\vec{A}$ for which $l = 1, m = 1, 0 - 1$ can be found from the Wigner–Eckart theorem. A calculation yields:

$$\begin{aligned}
\langle lm|A_3|lm\rangle &= \frac{m}{\sqrt{l(l+1)(2l+1)}}\langle l||A||l\rangle \\
\langle lm|A_3|l-1, m\rangle &= \sqrt{\frac{l^2 - m^2}{l(2l-1)(2l+1)}}\langle l||A||l-1\rangle \\
\langle l-1, m|A_3|lm\rangle &= \sqrt{\frac{l^2 - m^2}{l(2l-1)(2l+1)}}\langle l-1||A||l\rangle \\
\langle l, m-1|A_-|lm\rangle &= \sqrt{\frac{(l-m+1)(l+m)}{l(2l+1)(2l+1)}}\langle l||A||l\rangle
\end{aligned} \tag{4.62}$$

$$\langle l, m-1|A_-|l-1m\rangle = \sqrt{\frac{(l-m+1)(l-m)}{l(2l-1)(2l+1)}}\langle l||A||l-\rangle \tag{4.63}$$

$$\langle l-1, m-1|A_-|lm\rangle = -\sqrt{\frac{(l+m-1)(l+m)}{l(2l-1)(2l+1)}}\langle l-1||A||l\rangle$$

Matrix elements of A_+ can be obtained from those of A_- by using the relation

$$\langle L'M'|A_+|LM\rangle = \langle LM|A_-|L'M'\rangle^*.$$

Problems

1. Show that

$$\langle l||L||l\rangle = \sqrt{l(l+1)(2l+1)},$$
$$\langle l-1||L||l\rangle = \langle l||L||l-1\rangle = 0.$$

2. Show by calculation of matrix elements of $n_z = \cos\theta$

$$\langle l-1, 0|\cos\theta|l, 0\rangle$$

that $\langle l||n||l\rangle = \sqrt{l}$

$$\langle l||n||l\rangle = 0.$$

3. Using (4.5), show that for a free particle with $H = \frac{p^2}{2m}$ the commutator of $x(t)$ with $x(t')$ is

$$[x(t), x(t')] = \frac{i\hbar}{m}(t-t').$$

4. Given a wavefunction

$$\psi(\theta,\varphi) = \begin{cases} 1 & \text{for } 0 < \theta < \frac{\pi}{2} \\ & \quad\; 0 < \varphi < \frac{\pi}{2} \\ 0 & \text{elsewhere} \end{cases}$$

on the surface of a sphere, expand it in spherical harmonics and evaluate it at $\theta = \pi/4$, $\varphi = \pi/4$ and at $\theta = 3\pi/4$, $\varphi = \pi/4$.

5. On measurement it is found that

$$L^2 = 2\hbar,$$

$$L_z \cos\alpha + L_x \sin\alpha = 0.$$

Find the probability that the measurement of L_x will yield a value of 1.

Hint: $L^2 = 2\hbar$ means that L^2 and the components of $\vec{L}$ are 3×3 matrices. Write eigenvalue matrix equation for $L_z \cos\alpha + L_x \sin\alpha$. Use the eigenfunction as a basis and expand eigenfunction of L_x in this basis to obtain probabilities of its eigenvalues.

6. Three particles of spin $\frac{1}{2}$ each form a resultant $j = \frac{3}{2}$. Establish the relations

$$J_+|jm\rangle = \sqrt{j(j+1) - m(m+1)}\,|j, m+1\rangle$$

$$J_3|jm\rangle = m|jm\rangle$$

by using Pauli spin matrices of the individual spin-$\frac{1}{2}$ particles.
Hint: There are eight states with different $m_1 m_2$ each having values $\frac{1}{2}$ and $-\frac{1}{2}$ to form total $m = m_1 + m_2 + m_3 = \frac{3}{2}, \frac{1}{2}, -\frac{1}{2}, -\frac{3}{2}$. Obtain $\vec{J}^2(3\times 3)$ matrices for $m = \frac{1}{2}$ and $-\frac{1}{2}$ by using $\vec{J} = \frac{1}{2}(J_+J_- + J_-J_+) + J_3^2$ with $\vec{J} = \vec{S}_1 + \vec{S}_2 + \vec{S}_3$ and obtain their eigenvalues and eigenfunctions and thereby get $J_+|jm\rangle$ and $J_3|jm\rangle$. Do the same for $m = \frac{3}{2}$ and $-\frac{3}{2}$.

7. For a particle with $L = 2$ and $S = 1$, form the matrix $\vec{J}^2$ in the $|l, s, m_l, m_s\rangle$ representation for $m = m_l + m_s = 1$. Find the eigenvalues and eigenfunction of this matrix.
Hint: For $m = 1$, $|m_l m_s\rangle$ are $|2, -1\rangle$, $|1, 0\rangle$ and $|0, 1\rangle$. Obtain the $\vec{J}^2(3 \times 3)$ matrix by calculating J_+J_- J_-J_+ and J_z^2 matrices where $\vec{J} = \vec{L} + \vec{S}$ and then diagonalise the same to get eigenvalues and eigenfunctions.

8. Given for a spin-$\frac{1}{2}$ particle $\psi = \begin{pmatrix} a \\ b \end{pmatrix}$. Find the direction along which the spin will have eigenvalue 1 with 100% probability
Hint: For 2×2 matrix $\sigma = \alpha\sigma_x + \beta\sigma_y + \gamma\sigma_z$ whose eigenvalues are ± 1, obtain the eigenfunctions to get (α, β, γ)-direction for the desired probability.

Bibliography

[1] L.D. Landau and E.M. Lifshitz, *Quantum Mechanics*, 3rd edition, Pergamon Press, Oxford (1977).

[2] E.P. Wigner, *Gruppentheorie*, F. Vieweg (1931).

[3] G. Recah, *Phys. Rev.*, **62**, 146 (1942).

5 Quantum Mechanics of Photon and Neutrino

5.1 Introduction

We have seen in chapter 1 that Louis de Broglie postulated wave nature of material particles with wavelength inversely proportional to their momenta and Schrödinger gave a mathematical formulation of the dynamics of these waves by treating the classical dynamical variables of particles such as energy and momentum as differential operators. But a similar approach for the passage from light waves to photons is not given in most textbooks of quantum mechanics except a few, although many research publications have dealt with this topic in great detail. This is mainly because light waves are electromagnetic waves and passage to their particle nature is accomplished through quantization of the electromagnetic field. This field quantization goes by the name of second quantization while Schrödinger prescription is known as the first quantization. We shall present in this chapter Schrödinger's quantum mechanics for the photon. As a sequel, we shall also present the quantum mechanics of the neutrino.

5.2 Schrödinger Equation

Schrödinger equation

$$i\hbar\frac{\partial\psi}{\partial t} = -\frac{\hbar^2}{2m}\nabla^2\psi + V\psi \tag{5.1}$$

for de Broglie's matter waves for non-relativistic particles is obtained from the Newtonian dynamical relation

$$E = \frac{p^2}{2m} + V \tag{5.2}$$

through the prescription

$$\begin{aligned} E &\to i\hbar\frac{\partial}{\partial t}, \\ \vec{p} &\to -i\hbar\vec{\nabla}, \end{aligned} \tag{5.3}$$

by which E and $\vec{p}$ become differential operators. Substitution of (5.3) in (5.2) followed by operation on the wavefunction ψ gives Schrödinger equation (5.1). Its plane wave solutions for $V = 0$

$$\psi(\vec{r}, t) = \psi(0)e^{i\vec{k}\vec{r} - i\omega t} \tag{5.4}$$

give

$$\frac{\hbar^2 k^2}{2m} = \hbar\omega. \tag{5.5}$$

From (5.3) and (5.4) we find

$$\begin{aligned} p &= \hbar k = \hbar/\lambda, \\ E &= \hbar\omega, \end{aligned} \tag{5.6}$$

which are the de Broglie and the Einstein relations. These two relations connect the particle and the wave aspects of the Schrödinger wavefunction.

Since, classically, light propagates as waves which satisfy the Maxwell equations, the prescriptions to go over to the particle picture should be just the inverse of those for first quantization (5.3) of classical particles:

$$\begin{aligned} \vec{\nabla} &\to \frac{i}{\hbar}\vec{p}, \\ \frac{\partial}{\partial t} &\to \frac{-i}{\hbar}H. \end{aligned} \tag{5.7}$$

These prescriptions are to be used in the Maxwell equations which can be combined into a single one:

$$\partial_\nu G_{\mu\nu} = 0 \tag{5.8}$$

with

$$G_{\mu\nu} = F_{\mu\nu} + \bar{F}_{\mu\nu} \quad , \quad \bar{F}_{\mu\nu} - \frac{1}{2}\epsilon_{\mu\nu\lambda\eta}F_{\lambda\eta} \tag{5.9}$$

in the absence of charges and currents. In terms of ψ_i defined by

$$\begin{aligned} G_{ij} &= \epsilon_{ijk}(B_k - iE_k) = -i\epsilon_{ijk}\psi_k, \\ G_{i4} &= (B_i - iE_i) = -i\psi_i, \end{aligned} \tag{5.10}$$

Maxwell equation (5.8) takes the Schrödinger form

$$\begin{aligned} \frac{\partial \psi_i}{\partial t} &= -i\epsilon_{ijk}\partial_j\psi_k = -(S_j)_{ik}\partial_j\psi_k, \\ \partial_i\psi_i &= 0. \end{aligned} \tag{5.11}$$

On making use of the prescriptions (5.7) in the above equations, we get

$$\begin{aligned} H\psi_i &= -(S_j)_{ik} p_j \psi_k, \\ p_i \psi_i &= 0. \end{aligned} \tag{5.12}$$

The first of these two equations is the Schrödinger equation for the photon.

The neutrino equation is obtained by replacing spin-1 matrix $\vec{S}$ by spin-$\frac{1}{2}$ Pauli matrix $\vec{\sigma}$.

5.3 Photon with Definite Momentum

The Schrödinger equation (5.12), written explicitly has the form

$$E \begin{pmatrix} \psi_1 \\ \psi_2 \\ \psi_3 \end{pmatrix} = \begin{pmatrix} 0 & -ip_3 & +ip_2 \\ +ip_3 & 0 & -ip_1 \\ -ip_2 & +ip_1 & 0 \end{pmatrix} \begin{pmatrix} \psi_1 \\ \psi_2 \\ \psi_3 \end{pmatrix}, \tag{5.13}$$

where E is the energy eigenvalue

$$H\psi = E\psi. \tag{5.14}$$

The solution of this equation yields three eigenvalues:

$$E = \pm p, 0, \tag{5.15}$$

where

$$p = \sqrt{p_1^2 + p_2^2 + p_3^2}. \tag{5.16}$$

The corresponding orthonormal eigenfunctions are

$$\psi^{(+)} = \frac{1}{p\sqrt{2(p_1^2 + p_2^2)}} \begin{pmatrix} ipp_2 - p_1 p_3 \\ -ipp_1 - p_2 p_3 \\ p_1^2 + p_2^2 \end{pmatrix}, \tag{5.17a}$$

$$\psi^{(-)} = \frac{1}{p\sqrt{2(p_1^2 + p_2^2)}} \begin{pmatrix} -ipp_2 - p_1 p_3 \\ ipp_1 - p_2 p_3 \\ p_1^2 + p_2^2 \end{pmatrix}, \tag{5.17b}$$

$$\psi^{(0)} = \frac{1}{p} \begin{pmatrix} p_1 \\ p_2 \\ p_3 \end{pmatrix}. \tag{5.17c}$$

The third solution $\psi^{(0)}$ corresponding to zero energy does not satisfy the second equation of (5.12) and as such it has to be discarded. The solution $\psi^{(+)}$ has helicity

$$\lambda = \frac{\vec{S} \cdot \vec{p}}{p} = +1 \tag{5.18}$$

and corresponds to right-circularly-polarized photon. The negative energy solution $\psi^{(-)}$ has to be interpreted by writing the wave equation as

$$-\vec{S} \cdot \vec{p}\psi = p\psi \tag{5.19}$$

which tells us that it is a positive energy solution with negative helicity ($\lambda = -1$) which corresponds to left-circular polarization.

Equations (5.17) give us the photon wavefunction in the energy–momentum space. In configuration space,

$$\psi_i(\vec{r}) = i(B_i(\vec{r}) - iE_i(\vec{r})).$$

Its norm is not unity as can be seen from

$$\int d^3x\psi^+(\vec{r})\psi(\vec{r}) = \int d^3x(E^2 + B^2) = 2H. \tag{5.20}$$

One can construct a normalised wavefunction $\phi(\vec{r})$:

$$\phi(\vec{r}) = \frac{1}{\sqrt{2H}}\psi(\vec{r}) = \frac{1}{\sqrt{2H}}\int d^3pe^{i\vec{p}\cdot\vec{r}}\psi(\vec{p}). \tag{5.21}$$

But $\phi^+(\vec{r})\phi(\vec{r})$ cannot be taken as the probability of finding the photon at $\vec{r}$ because the measurables are

$$\psi_i(\vec{r}) = i(B_i - iE_i) \tag{5.22}$$

on account of which the probability of finding the photon at a given point in the configuration space is determined by the values of the wavefunction at all points. It is neverthless possible to obtain the expectation values of dynamical variables using this wavefunction.

5.4 Photon with Definite Angular Momentum

Electric and magnetic photons

Angular momentum operator J_i is the generator of infinitesimal rotation under which momentum transforms as

$$p_i \to p_i - \epsilon_{ijk}\theta_j p_k = p_i + i(S_j)_{ik}\theta_j p_k \tag{5.23}$$

and the photon eigenfunction $\psi_i(\vec{p})$ transforms as

$$\psi_i'(\vec{p}) = \psi_i(\vec{p}) + i(S_j)_{ik}\theta_j\psi_k(\vec{p}). \tag{5.24}$$

By Taylor's expansion,

$$\psi_i'(\vec{p}) = \psi_i'(\vec{p} + \delta\vec{p}) = \psi_i'(\vec{p}) + \delta p_k \partial_k \psi_i'(\vec{p}), \tag{5.25}$$

which on use in (5.24) gives

$$\begin{aligned}\delta\psi_i(\vec{p}) = \psi_i'(p) - \psi_i(\vec{p}) &= i(S_j)_{ik}\theta_j\psi_k(\vec{p}) + \vec{\theta}\cdot(\vec{p}\times\vec{\nabla}_p)_{ik}\psi_{ik}(\vec{p})\\ &= \vec{\theta}\cdot\vec{J}_{ij}\psi_j(\vec{p}),\end{aligned} \tag{5.26}$$

where

$$\vec{J} = \vec{p}\times\vec{\nabla}_p + \vec{S} = \vec{L} + \vec{S} \tag{5.27}$$

is the total angular momentum and $\vec{L}$ is the orbital angular momentum in the momentum space. It is easily verified that

$$[J_i, H] = 0, \tag{5.28}$$

which shows that the total angular momentum is conserved. It may be worth noting here that in the second quantized theory, the separation of the total angular momentum of the photon into orbital and spin parts is not unique. Since the vector potential appears in both, the separation is not gauge-invariant.

The simultaneous eigenfunctions ψ_{jlm} of J^2, L^2, S^2, J_3, L_3 and S_3 as listed below.

$$\begin{aligned}J^2\psi_{jlm} &= j(j+1)\psi_{jlm}\\ J_3\psi_{jlm} &= m\psi_{jlm}\\ L^2\psi_{jlm} &= l(l+1)\psi_{jlm}\\ L_3\psi_{jlm} &= m_l\psi_{jlm}\\ S^2\psi_{jlm} &= 2\psi_{jlm}\\ S_3\psi_{jlm} &= m_s\psi_{jlm}\end{aligned} \tag{5.29}$$

$j = l; l \pm 1$ and $m = m_l + m_s$ are obtained as follows.
We shall use the representation for S_i:

$$S_3 = \begin{pmatrix}1 & 0 & 0\\ 0 & 0 & 0\\ 0 & 1 & -1\end{pmatrix},\quad S_1 = \frac{1}{\sqrt{2}}\begin{pmatrix}0 & 1 & 0\\ 1 & 0 & 1\\ 0 & 1 & 0\end{pmatrix},\quad S_2 = \frac{1}{\sqrt{2}}\begin{pmatrix}0 & -i & 0\\ i & 0 & -i\\ 0 & i & 0\end{pmatrix} \tag{5.30}$$

so that

$$J_3 = \begin{pmatrix} 1 - i\frac{\partial}{\partial\varphi} & 0 & 0 \\ 0 & -i\frac{\partial}{\partial\varphi} & 0 \\ 0 & 0 & -1 - i\frac{\partial}{\partial\varphi} \end{pmatrix}. \tag{5.31}$$

The equations $J_3\psi = m\psi$ and $L^2\psi = l(l+1)\psi$ when solved give

$$\psi = \begin{pmatrix} f & Y_{l,m-1}(\theta,\varphi) \\ g & Y_{l,m}(\theta,\varphi) \\ h & Y_{l,m+1}(\theta,\varphi) \end{pmatrix}, \tag{5.32}$$

where f, g and h are to be determined from equations (5.29). Using this in the equation

$$J^2\psi = \begin{pmatrix} L^2 + S^2 + 2L_3 & \sqrt{2}L^{(-)} & 0 \\ \sqrt{2}L^{(+)} & L^2 + S^2 & \sqrt{2}L^{(-)} \\ 0 & \sqrt{2}L^{(+)} & L^2 + S^2 - 2L_3 \end{pmatrix} \begin{pmatrix} f & Y_{l,m-1}(\theta,\varphi) \\ g & Y_{l,m}(\theta,\varphi) \\ h & Y_{l,m+1}(\theta,\varphi) \end{pmatrix}$$

$$= j(j+1) \begin{pmatrix} f & Y_{l,m-1}(\theta,\varphi) \\ g & Y_{l,m}(\theta,\varphi) \\ h & Y_{l,m+1}(\theta,\varphi) \end{pmatrix}, \tag{5.33}$$

one gets

$$\begin{aligned} &[l(l+1) + 2m - j(j+1)]f + \sqrt{2(l+m)(l-m+1)}\, g = 0, \\ &[l(l+1) - 2m - j(j+1)]h + \sqrt{2(l-m)(l+m+1)}\, g = 0. \end{aligned} \tag{5.34}$$

One further relation that has to be used is the normalisation condition

$$f^2 + g^2 + h^2 = 1. \tag{5.35}$$

After some simple algebraic manipulations, one obtains for the cases $j = l, l-1$ and $l+1$, the following eigenfunctions

$$\psi_{jlm}(\theta\varphi) = \begin{pmatrix} -\sqrt{\frac{(j+1)(j-m+1)}{2j(j+1)}} & Y_{j,m-1}(\theta,\varphi) \\ \sqrt{\frac{m^2}{j(j+1)}} & Y_{jm}(\theta,\varphi) \\ \sqrt{\frac{(j-m)(j+m+1)}{2j(j+1)}} & Y_{j,m+1}(\theta,\varphi) \end{pmatrix} \tag{5.36}$$

which has parity

$$P\psi_{jjm}(\theta,\varphi) = (-)^{j+1}\psi_{jjm}(\theta,\varphi). \tag{5.37}$$

$$\psi_{j,j+1,m}(\theta,\varphi) = \begin{pmatrix} +\sqrt{\frac{(j-m+2)(j-m+1)}{(2j+2)(2j+3)}} & Y_{j+1,m-1}(\theta,\varphi) \\ -\sqrt{\frac{(j+m+1)(j-m+1)}{(j+1)(2j+3)}} & Y_{j+1,m}(\theta,\varphi) \\ +\sqrt{\frac{(j+m+2)(j+m+1)}{(2j+2)(2j+3)}} & Y_{j+1,m+1}(\theta,\varphi) \end{pmatrix} \tag{5.38}$$

which has parity

$$P\psi_{j,j+1,m}(\theta,\varphi) = (-1)^j\psi_{j,j+1,m}(\theta,\varphi). \tag{5.39}$$

$$\psi_{j,j-1,m}(\theta,\varphi) = \begin{pmatrix} \sqrt{\frac{(j+m)(j+m-1)}{2j(2j-1)}} & Y_{j-1,m-1}(\theta,\varphi) \\ \sqrt{\frac{j^2-m^2}{j(2j-1)}} & Y_{j-1,m}(\theta,\varphi) \\ \sqrt{\frac{(j-m)(j-m-1)}{2j(2j-1)}} & Y_{j-1,m+1}(\theta,\varphi) \end{pmatrix} \tag{5.40}$$

which has parity

$$P\psi_{j,j-1,m}(\theta,\varphi) = (-1)^j\psi_{j,j-1,m}(\theta,\varphi). \tag{5.41}$$

Each of these are vectors and known as vector spherical harmonics. A linear combination

$$Y^{(e)}_{jm}(\theta,\varphi) = \sqrt{\frac{j}{2j+1}}\,\psi_{j,j+1,m}(\theta,\varphi) + \sqrt{\frac{j+1}{2j+1}}\,\psi_{j,j-1,m}(\theta,\varphi) \tag{5.42}$$

with parity $(-1)^j$ is called the 'electric photon' E_j eigenfunction while

$$Y^{(m)}_{jm}(\theta,\varphi) = \psi_{j,j,m}(\theta,\varphi) \tag{5.43}$$

with parity $(-1)^{j+1}$ is called the 'magnetic photon' M_j eigenfunction. Since the linear combination

$$Y^{(l)}_{jm}(\theta,\varphi) = \sqrt{\frac{j}{2j+1}}\,\psi_{j,j+1,m}(\theta,\varphi) - \sqrt{\frac{j+1}{2j+1}}\,\psi_{j,j-1,m}(\theta,\varphi) \tag{5.44}$$

orthogonal to $Y^{(e)}_{jm}(\theta,\varphi)$ with parity $(-1)^j$ happens to be longitudinal, it does not represent photons. These three spherical harmonics vectors are related to the scalar spherical harmonics by

$$\vec{Y}^{(e)}_{jm}(\theta,\varphi) = \frac{1}{\sqrt{j(j+1)}}\vec{\nabla}Y_{jm}(\theta,\varphi),$$
$$\vec{Y}^{(m)}_{jm}(\theta,\varphi) = \frac{1}{\sqrt{j(j+1)}}(\vec{n}\times\vec{\nabla}Y_{jm}(\theta,\varphi)), \tag{5.45}$$
$$\vec{Y}^{(l)}_{jm}(\theta,\varphi) = \vec{n}Y_{jm}(\theta,\varphi),$$

where $\vec{n} = \vec{p}/|\vec{p}|$ is the unit vector along $\vec{p}$. On simplification, these take the form

$$\vec{Y}^{(e)}_{jm}(\theta,\varphi) = \frac{1}{\sqrt{j(j+1)}}\begin{pmatrix} 0 \\ \frac{\partial}{\partial\theta}[Y_{jm}(\theta,\varphi)] \\ \frac{1}{\sin\theta}\frac{\partial}{\partial\phi}[Y_{jm}(\theta,\varphi)] \end{pmatrix}, \tag{5.46}$$

$$\vec{Y}^{(m)}_{jm}(\theta,\varphi) = \frac{1}{\sqrt{j(j+1)}}\begin{pmatrix} 0 \\ -\frac{1}{\sin\theta}\frac{\partial}{\partial\phi}[Y_{jm}(\theta,\varphi)] \\ \frac{\partial}{\partial\theta}[Y_{jm}(\theta,\varphi)] \end{pmatrix}, \quad \vec{Y}^{(l)}_{jm}(\theta,\varphi) = \begin{pmatrix} Y_{jm}(\theta,\varphi) \\ 0 \\ 0 \end{pmatrix}. \tag{5.47}$$

5.5 Photon Polarisation

Classical electromagnetic waves, because of their vectorial nature, have polarisation. The photon, too, has polarisation on account of its spin. Its wavefunction $\vec{\psi}$ can be written in terms of the polarisation vector $\vec{e}$.

$$\vec{\psi}(\vec{p}) = \vec{e}u(\vec{p}) \tag{5.48}$$

with

$$\vec{p}\cdot\vec{e} = 0.$$

If we take $\vec{p} = (0, 0, p)$, the polarisation vector can be represented by

$$\vec{e} = \vec{e}_1\cos\alpha + \vec{e}_2\sin\alpha e^{\iota\beta}. \tag{5.49}$$

When the photon is interacting with other particles such as electrons, it cannot be represented by a wavefunction. It can, however, be represented by the density matrix $\rho_{\alpha,\beta}$ defined by

$$\rho_{\alpha,\beta} = \overline{e_\alpha e^*_\beta}, \quad \alpha,\beta = 1, 2, \tag{5.50}$$

where the bar represents averaging over interaction parameters. From this definition it follows that

$$\rho_{\alpha,\beta} = \rho^*_{\beta,\alpha}, \tag{5.51}$$

i.e., the density matrix is hermitian. From the normalisation condition

$$|e_1|^2 + e_2|^2 = 1, \tag{5.52}$$

it follows that

$$\mathrm{Tr}\,\rho = \rho_{11} + \rho_{22} = 1. \tag{5.53}$$

Such a hermitian matrix whose trace is unity, can in general be written in terms of three real parameters, ξ_1, ξ_2, ξ_3, known as Stokes parameters in the form

$$\rho = \frac{1}{2}\begin{pmatrix} 1+\xi_3 & \xi_1 - i\xi_2 \\ \xi_1 + i\xi_2 & 1-\xi_3 \end{pmatrix} = \frac{1}{2}(1 + \vec{\sigma}\cdot\vec{\xi}). \tag{5.54}$$

If we use

$$\vec{e}_1 = (1\ ,\ 0)\ ,\ \vec{e}_2 = (0\ ,\ 1) \tag{5.55}$$

in (5.49), we get

$$\rho = \begin{pmatrix} \cos^2\alpha & \frac{1}{2}\sin 2\alpha e^{-i\beta} \\ \frac{1}{2}\sin 2\alpha e^{i\beta} & \sin^2\alpha \end{pmatrix}. \tag{5.56}$$

Comparing with (5.54), the Stokes parameters turn out to be

$$\begin{aligned} \xi_1 &= \sin 2\alpha \cos\beta, \\ \xi_2 &= \sin 2\alpha \sin\beta, \\ \xi_3 &= \cos 2\alpha, \end{aligned} \tag{5.57}$$

so that

$$\xi_1^2 + \xi_2^2 + \xi_3^2 = 1, \tag{5.58}$$

which can be taken as prescription for complete polarisation. Unpolarised photon will be characterised by

$$\xi_1 = \xi_2 = \xi_3 = 0. \tag{5.59}$$

For partially polarised light,

$$\xi_i = p\eta_i, \qquad \eta_1^2 + \eta_2^2 + \eta_3^2 = 1, \tag{5.60}$$

where p is degree of polarisation. For such photon the density matrix is

$$\rho = \frac{1}{2}(1-p) + \frac{1}{2}(p + \vec{\eta}\cdot\vec{\sigma}). \tag{5.61}$$

5.6 Neutrino with Definite Momentum

The Schrödinger equation for neutrino is obtained by replacing the spin-1 matrices in the photon equation by spin-$\frac{1}{2}$ Pauli matrices.

$$-\vec{\sigma}_{\alpha\beta} \cdot \vec{p}\psi_{\beta} = \epsilon\psi_{\alpha}, \tag{5.62}$$

where

$$\sigma_1 = \begin{pmatrix} 0 & 1 \\ 1 & 0 \end{pmatrix}, \sigma_2 = \begin{pmatrix} 0 & -i \\ i & 0 \end{pmatrix} \quad \text{and} \quad \sigma_3 = \begin{pmatrix} 1 & 0 \\ 0 & -1 \end{pmatrix} \tag{5.63}$$

are the Pauli spin matrices. Substituting (5.63) in (5.62) gives

$$\begin{pmatrix} p_3 & p_1 - ip_2 \\ p_1 + ip_2 & -p_3 \end{pmatrix} \begin{pmatrix} \psi_1 \\ \psi_2 \end{pmatrix} = -\epsilon \begin{pmatrix} \psi_1 \\ \psi_2 \end{pmatrix}. \tag{5.64}$$

The energy eigenvalues obtained by the solution of this matrix equation are

$$\epsilon = \pm\sqrt{p_1^2 + p_2^2 + p_3^2} = \pm p. \tag{5.65}$$

Substituting $\epsilon = p$ in (5.64) and using normalisation condition

$$|\psi_1|^2 + |\psi_2|^2 = 1 \tag{5.66}$$

gives

$$\psi^{(+)} = \left(\frac{p + p_3}{2p}\right)^{1/2} \begin{pmatrix} 1 \\ \frac{p-p_3}{p_1-ip_2} \end{pmatrix}. \tag{5.67}$$

For $\epsilon = -p$, one gets

$$\psi^{(-)} = \left(\frac{p - p_3}{2p}\right)^{1/2} \begin{pmatrix} 1 \\ \frac{-(p_3 + p)}{p_1-ip_2} \end{pmatrix}. \tag{5.68}$$

As in the photon case, the helicities of $\psi^{(\pm)}$ are $\pm 1/2$:

$$\lambda = \frac{\vec{S} \cdot \vec{p}}{p} = \frac{1}{2}\frac{\vec{\sigma} \cdot \vec{p}}{p}\psi^{(\pm)} = \pm\frac{1}{2}\psi^{(\pm)} \tag{5.69}$$

with the negative energy solution as the one with positive energy and negative helicity.

5.7 Neutrino with Definite Angular Momentum

Under infinitesimal rotation where the neutrino momentum transforms as

$$p_i \to p_i + \epsilon_{ijk}\theta_j p_k, \tag{5.70}$$

its wavefunction will transform as

$$\psi'_\alpha(\vec{p}\,') = \psi_\alpha(\vec{p}) + \frac{1}{2}\vec{\sigma}_{\alpha\beta}\cdot\vec{\theta}\psi_\beta(\vec{p}). \tag{5.71}$$

As in the photon case,

$$\psi'_\alpha(\vec{p}\,') = \psi'_\alpha(\vec{p}+\delta\vec{p}) = \psi'_\alpha(\vec{p}) + \cdot\delta p_i\partial_i\psi'_\alpha(\vec{p}) \tag{5.72}$$

so that

$$\delta\psi_\alpha(\vec{p}) = \psi'_\alpha(\vec{p}) - \psi_\alpha(\vec{p}) = \frac{1}{2}\vec{\sigma}_{\alpha\beta}\cdot\vec{\theta}\psi_\beta(\vec{p}) + \vec{\theta}\cdot(\vec{p}\times\vec{\nabla}_p)\psi_\alpha(\vec{p}). \tag{5.73}$$

Since

$$\delta\psi_\alpha(\vec{p}) = \vec{\theta}\cdot\left(\vec{L}\delta_{\alpha\beta} + \frac{1}{2}\vec{\sigma}_{\alpha\beta}\right)\psi_\beta(\vec{p}) = \vec{\theta}\cdot\vec{J}_{\alpha\beta}\psi_\beta \tag{5.74}$$

where

$$\vec{J} = \vec{L} + \frac{1}{2}\vec{\sigma} = \vec{L} + \vec{S}$$

is the total angular momentum in the momentum space which commutes with the Hamiltonian $H = \vec{\sigma}\cdot\vec{p}$:

$$[H, J_i] = 0 \tag{5.75}$$

as in the case of photon. For determining the simultaneous eigenfunctions of the commuting operators $J^2, L^2, S^2, J_3, L_3, S_3$, we proceed as in the photon case. Here

$$J_3 = L_3 + S_3 = \begin{pmatrix} \left(\frac{1}{2} - i\frac{\partial}{\partial\varphi}\right) & 0 \\ 0 & \left(-\frac{1}{2} - i\frac{\partial}{\partial\varphi}\right) \end{pmatrix}, \tag{5.76}$$

$$\text{and } J_3\psi = m\psi \tag{5.77}$$

gives

$$\psi \sim \begin{pmatrix} e^{i(m-\frac{1}{2})\varphi} \\ e^{i(m+\frac{1}{2})\varphi} \end{pmatrix}. \tag{5.78}$$

We therefore set

$$\psi = \begin{pmatrix} f\ Y_{l,m-\frac{1}{2}}(\theta\varphi) \\ g\ Y_{l,m+\frac{1}{2}}(\theta\varphi) \end{pmatrix} \tag{5.79}$$

and substitute it in

$$J^2\psi = \begin{pmatrix} L^2 + \frac{3}{4} + L_3 & L^{(-)} \\ L^{(+)} & L^2 + \frac{3}{4} - L_3 \end{pmatrix} \begin{pmatrix} \psi_1 \\ \psi_2 \end{pmatrix} = j(j+1) \begin{pmatrix} \psi_1 \\ \psi_2 \end{pmatrix} \tag{5.80}$$

to obtain

$$\left[l(l+1) + \frac{1}{4} + m\right] f + \sqrt{\left(l + m + \frac{1}{2}\right)\left(l - m + \frac{1}{2}\right)}\, g = j(j+1) f, \tag{5.81a}$$

$$\sqrt{\left(l + m + \frac{1}{2}\right)\left(l - m + \frac{1}{2}\right)}\, f + \left[l(l+1) + \frac{1}{4} - m\right] g = j(j+1) g. \tag{5.81b}$$

In addition, f and g are related by normalisation of the wavefunction

$$f^2 + g^2 = 1. \tag{5.81c}$$

The rules of addition of the angular momentum tells us that j can either be $l + \frac{1}{2}$ or $l - \frac{1}{2}$. For $j = l + \frac{1}{2}$, (5.81) gives

$$f = \frac{\sqrt{j+m}}{2j}, \qquad g = \frac{\sqrt{j-m}}{2j} \tag{5.82}$$

so that

$$\psi_{j,j-\frac{1}{2},m}(\vec{p}) = \begin{pmatrix} \sqrt{\frac{j+m}{2j}} & Y_{j-\frac{1}{2},m-\frac{1}{2}}(\theta\varphi) \\ \sqrt{\frac{j-m}{2j}} & Y_{j-\frac{1}{2},m+\frac{1}{2}}(\theta\varphi) \end{pmatrix}. \tag{5.83}$$

For $j = l - \frac{1}{2}$, the same equations give

$$f = -\sqrt{\frac{j-m+1}{2j+2}}, \qquad g = \sqrt{\frac{j+m+1}{2j+2}}, \tag{5.84}$$

so that

$$\psi_{j,j+\frac{1}{2},m}(\vec{p}) = \begin{pmatrix} -\sqrt{\frac{j+m}{2j+2}} & Y_{j+\frac{1}{2},m-\frac{1}{2}}(\theta\varphi) \\ \sqrt{\frac{j+m+1}{2j+2}} & Y_{j+\frac{1}{2},m+\frac{1}{2}}(\theta\varphi) \end{pmatrix}. \tag{5.85}$$

The angular momentum wavefunction (5.83) and (5.85) are known as spinor spherical harmonics.

Problems

1. Show that photon is a composite of neutrino and its anti particle, i.e.,

$$\psi_i(\vec{p}) = \tilde{\xi}_\alpha(\vec{p}^{(2)})\sigma_i^{\alpha\beta}\eta_\beta(\vec{p}^{(1)}),$$

where $\psi_i(\vec{p})$ is photon wavefunction, η_α is the wavefunction of the neutrino and ξ_α that of the anti neutrino satisfying the equations

$$-i\hbar\frac{\partial\psi}{\partial t} = \vec{S}\cdot\vec{p}\psi,$$

$$-i\hbar\frac{\partial\eta}{\partial t} = -\vec{\sigma}\cdot\vec{p}\eta,$$

$$-i\hbar\frac{\partial\xi}{\partial t} = \vec{\sigma}\cdot\vec{p}\xi,$$

where η and ξ are related by

$$\eta^*(-\vec{p}) = \xi(\vec{p})$$

and

$$\vec{p} = \vec{p}^{(1)} + \vec{p}^{(2)}.$$

2. By Schrödinger quantization of classical equations

$$\partial_i E_{ij} = 0 \quad \partial_i B_{ij} = 0$$
$$\mathcal{E}_{ikl}\partial_l E_{kj} + \frac{\partial B_{ij}}{\partial t} = 0$$
$$\mathcal{E}_{ikl}\partial_l B_{kj} - \frac{\partial E_{ij}}{\partial t} = 0$$
$$E_{ij} = E_{ji}, \quad B_{ij} = B_{ji}$$
$$E_{11} + E_{22} + E_{33} = 0$$
$$B_{11} + B_{22} + B_{33} = 0,$$

find out what spin the quantum mechanical equations represent.

3. By actual calculations, decide which of the following operators is hermitian.

$$(a)\ \vec{r}\times\vec{L} \quad (b)\ (\vec{L}\times\vec{r} + \vec{r}\times\vec{L}) \quad (c)\ (\vec{L}\times\vec{r} - \vec{r}\times\vec{L})$$

Bibliography

[1] A.I. Akhiezer and V.B. Berestetskii, *Quantum Electrodynamics*, Interscience Publisher (1965).

[2] B. Korsunoglu, *Modern Quantum Theory*, W.H. Freeman and Company, London (1962).

[3] J.R. Oppenheimer, *Phys. Rev.*, **38**, 725 (1931).

[4] W.J. Archibald, *Can. J. Phys.*, **33**, 565 (1955).

[5] R.H. Good, *Phys. Rev.*, **105**, 1914 (1957).

[6] T. Pradhan, *The Photon*, Nova Science Publishers, New York (2001).

6 Passage from Quantum to Classical Mechanics

6.1 Introduction

In classical mechanics, particles have well defined paths in contrast to quantum mechanics where these particles are represented by waves which obey Schrödinger equation. The situation is similar to geometrical optics where light moves in well defined paths called rays in contrast to wave optics where motion of light is determined by wave equation. It is well-known that in the limit of short wavelengths, one obtains geometrical optics from wave optics. This is the zeroeth order in the approximation scheme known as eikonal approximation. The classical limit of wave mechanics is obtained in a similar manner. The Planck's constant plays the role of wavelength. The approximation scheme in this case was developed independently by Wentzel, Kramers and Brillouin and is popularly known as WKB approximation.

In this chapter we shall first present the optical case and by analogy obtain the classical limit of wave mechanics. The only difference between the two is that while optical wave equation is second order in time, the quantum mechanical equation is first order in time with an imaginary coefficient.

6.2 Geometrical Optics from Wave Optics

Geometrical optics is the small wavelength limit of wave optics. In this limiting case the wave can be represented as

$$\vec{E}(\vec{r},t) = \vec{E}_0(\vec{r},t)e^{\frac{i}{\lambda}\phi(\vec{r},t)}, \tag{6.1}$$

where the wavelength is small compared to the dimension of the space in which $\vec{E}$ is appreciable. Substitution of this representation in the wave equation

$$\nabla^2\vec{E} - \frac{1}{v^2}\frac{\partial^2\vec{E}}{\partial t^2} = 0,$$

v being the phase velocity, yields

$$\left(\nabla^2 - \frac{1}{v^2}\frac{\partial^2}{\partial t^2}\right)\vec{E}_0 - \frac{\vec{E}_0}{\lambda^2}\left[(\nabla\phi)^2 - \frac{1}{v^2}\left(\frac{\partial \phi}{\partial t}\right)^2\right]$$
$$+\frac{i\vec{E}_0}{\lambda}\left(\nabla^2\phi - \frac{1}{v^2}\frac{\partial^2\phi}{\partial t^2}\right) + \frac{2i}{\lambda}\left(\frac{\partial \vec{E}_0}{\partial x_i}\frac{\partial \phi}{\partial x_i} - \frac{1}{v^2}\frac{\partial \vec{E}_0}{\partial t}\frac{\partial \phi}{\partial t}\right) = 0. \quad (6.2)$$

Equating to zero separately the real and imaginary parts of this equation gives,

$$\vec{E}_0\left[(\vec{\nabla}\phi)^2 - \frac{1}{v^2}\left(\frac{\partial \phi}{\partial t}\right)^2\right] = \lambda^2\left(\nabla^2 - \frac{1}{v^2}\frac{\partial^2}{\partial t^2}\right)\vec{E}_0 \quad (6.3)$$

$$\frac{\partial}{\partial x_i}\left(E_0^2\frac{\partial \phi}{\partial x_i}\right) - \frac{1}{v^2}\frac{\partial}{\partial t}\left(E_0^2\frac{\partial \phi}{\partial t}\right) = 0. \quad (6.4)$$

While (6.4) is a continuity equation, (6.3) in the limit $\lambda = 0$ reduces to

$$(\vec{\nabla}\phi)^2 - \frac{1}{v^2}\left(\frac{\partial \phi}{\partial t}\right)^2 = 0, \quad (6.5)$$

which is a first order differential equation. Since in the limit of a plane wave $\phi/\lambda = \vec{k}\cdot\vec{r} - \omega t$, (6.5) reduces to

$$k^2 - \frac{\omega^2}{v^2} = 0. \quad (6.6)$$

We shall define, in the general case, $\vec{k} = \vec{\nabla}\phi, \quad \omega = -\frac{\partial\phi}{\partial t}$.
Equation (6.5) then reduces to

$$(\vec{\nabla}\phi)^2 = -n^2, \quad (6.7)$$

which is the eikonal equation of geometrical optics according to which the wave vector $\vec{k}$ is normal to the surface of constant phase $\phi = \phi_0$ representing the direction of light rays.

6.3 Classical Mechanics from Wave Mechanics

The passage from wave mechanics to classical mechanics is obtained in the same way as that of wave optics to geometrical optics. In place of the optical wave equation which is second order in time, the Schrödinger equation is first order in time and also has an imaginary coefficient.

$$i\hbar\frac{\partial \psi}{\partial t} - \frac{\hbar^2}{2m}\nabla^2\psi - V\psi = 0. \quad (6.8)$$

In order to pass to the classical limit, we follow the procedure of the previous section and represent ψ as

$$\psi(x,t) = \psi_0(x,t)e^{\frac{iS(x,t)}{\hbar}} \tag{6.9}$$

with ψ_0 and S as real function for systems where the de Broglie wavelength of particle is small compared to the dimensions of the system and then substitute it in the Schrödinger equation (6.8). This gives us

$$\psi_0\frac{\partial S}{\partial t} - i\hbar\frac{\partial \psi_0}{\partial t} + \frac{\psi_0}{2m}(\vec{\nabla}S)^2 - \frac{i\hbar}{2m}\psi_0\nabla^2 S - \frac{i\hbar}{m}(\vec{\nabla}S)\cdot(\vec{\nabla}\psi_0)$$

$$-\frac{\hbar^2}{2m}\nabla^2\psi_0 + V\psi_0 = 0, \tag{6.10}$$

from which we get, on equating separately the real and imaginary parts to zero,

$$\frac{\partial S}{\partial t} + \frac{1}{2m}(\vec{\nabla}S)^2 + V - \frac{\hbar^2}{2m}\cdot\frac{1}{\psi_0}\nabla^2\psi_0 = 0, \tag{6.11}$$

$$\frac{\partial \psi_0}{\partial t} + \frac{\psi_0}{2m}\nabla^2 S + \frac{1}{m}(\vec{\nabla}S)\cdot(\vec{\nabla}\psi_0) = 0. \tag{6.12}$$

In the limit $\hbar = 0$, the first equation becomes the Hamilton–Jacobi (H–J) equation of classical mechanics. The second equation when multiplied by ψ_0 becomes a continuity equation

$$\frac{\partial}{\partial t}(\psi_0^2) + \vec{\nabla}\cdot\left(\psi_0^2\frac{\vec{\nabla}S}{m}\right) = 0. \tag{6.13}$$

Since, in classical mechanics $\vec{p} = \vec{\nabla}S$, $\psi_0^2\frac{\vec{\nabla}S}{m} = \psi_0^2\vec{v}$, where $\vec{v}$ is the velocity, it is a relation between probability density ψ_0^2 and probability current density $\psi_0^2\vec{v}$. It is worth noting that even though S obeys classical H–J equation, the system still has a wavefunction obeying wave mechanical equation (6.8). Hence we have a quasi-classical system represented by (6.9) and the H–J equation. It may be noted that Schrödinger started with $S = K\ln\psi$ (Appendix A), which is equivalent to (6.9) and used it in the Hamilton–Jacobi equation to obtain his wave equation. Here we have done just the reverse.

6.4 WKB Approximation

If we take ψ_0 of the previous section to be a constant independent of space time and use the identity

$$\frac{\partial S}{\partial t} = -E, \tag{6.14}$$

then (6.10) reduces to

$$\frac{1}{2m}(\vec{\nabla}S)^2 - \frac{i\hbar}{2m}\nabla^2 S = E - V. \tag{6.15}$$

Since we have not separated the real and immaginary parts, S here is complex. We now expand S in a power series in $\hbar/i$

$$S = S_0 + \frac{\hbar}{i}S_1 + \left(\frac{\hbar}{i}\right)^2 S_2 + \cdots, \tag{6.16}$$

and substitute the same in (6.15) and equate the terms independent of $\hbar$ and then of the first power of $\hbar$ to get

$$\left(\vec{\nabla}S_0\right)^2 = 2m(E - V) \tag{6.17}$$

$$\frac{1}{2}\nabla^2 S_0 + \vec{\nabla}S_0 \cdot \vec{\nabla}S_1 = 0. \tag{6.18}$$

For one-dimensional systems these equations are

$$\left(\frac{dS_0}{dx}\right)^2 = 2m(E - V), \tag{6.19}$$

$$\frac{1}{2}\frac{d^2S_0}{dx^2} + \frac{dS_0}{dx}\frac{dS_1}{dx} = 0. \tag{6.20}$$

Solution of (6.19) is

$$S_0 = \pm \int dx\sqrt{2m(E - V)} = \pm \int dx p(x) \tag{6.21}$$

while that of (6.20) is

$$S_1 = -\frac{1}{2}\log p. \tag{6.22}$$

Substitution of (6.16), (6.21) and (6.22) in (6.9) gives

$$\psi = \psi_0 e^{iS_0/\hbar + S_1} = \psi_0 \frac{e^{\pm\frac{i}{\hbar}\int dx p(x)}}{\sqrt{p}}. \tag{6.23}$$

In general,

$$\psi = C^{(+)}p^{-1/2}e^{+\frac{i}{\hbar}\int dx p(x)} + C^{(-)}p^{-1/2}e^{-\frac{i}{\hbar}\int dx p(x)}, \tag{6.24}$$

where $C^{(+)}$ and $C^{(-)}$ are constants. It will be seen that the probability of finding the particle within a segment dx is proportional to $\frac{1}{p} \sim \frac{1}{v}$, which is the time spent by the classical particle in that segment. For regions of space where $E > V$, it is convenient to put (6.23) into the form

$$\psi(x) = \frac{A}{\sqrt{p}} \sin\left(\int dxp + \theta\right), \tag{6.25}$$

and for region $V > E$ which is classically inaccessible,

$$\psi(x) = \frac{B}{\sqrt{p}} \exp\left(\pm\frac{1}{\hbar}\int dxp + \theta\right). \tag{6.26}$$

At this stage, let us find out the condition of validity of this approximation. It is clear from (6.15) that it is valid if the magnitude of the second term in it is much less than that of the first, i.e.,

$$\hbar\left|\frac{d^2S/dx^2}{(dS/dx)^2}\right| = \left|\hbar\frac{d}{dx}\left(\frac{1}{\frac{dS}{dx}}\right)\right| \ll 1. \tag{6.27}$$

Since $dS/dx = p(x) = h/\lambda$, this condition boils down to

$$\left|\frac{d\lambda}{dx}\right| \ll 1. \tag{6.28}$$

Physically, this condition states that the variation of λ within a distance λ should be very small. Since at the points where $E = V(x)$ (i.e., at turning points $p \to 0$ and $\lambda \to \infty$) the variation of $p \sim \lambda$ is very large, this approximation is inapplicable. Near such a turning point,

$$k^2 = \frac{2m}{\hbar^2}(E - V) = xa^2, \tag{6.29}$$

where a^2 = constant. The Schrödinger equation

$$\frac{d^2\psi}{dx^2} + a^2x\psi = 0 \tag{6.30}$$

for this case has the solution

$$\psi(x) = x^{1/2}\left[C_1J_{1/3}\left(\frac{2}{3}ax^{3/2}\right) + C_2N_{1/3}\left(\frac{2}{3}ax^{3/2}\right)\right], \tag{6.31}$$

where $J_{1/3}$ and $N_{1/3}$ are Bessel and Neuman function and this solution has asymptotic values:

$$\lim_{x\to+\infty}\psi(x) = \frac{1}{x^{1/4}}\sin\left(\frac{2x^{3/2}}{3}a + \frac{\pi}{4}\right),$$

$$\lim_{x\to-\infty}\psi(x) = \frac{1}{2(-x)^{1/4}}\exp\left[-\frac{2}{3}a(-x)^{3/2}\right]. \tag{6.32}$$

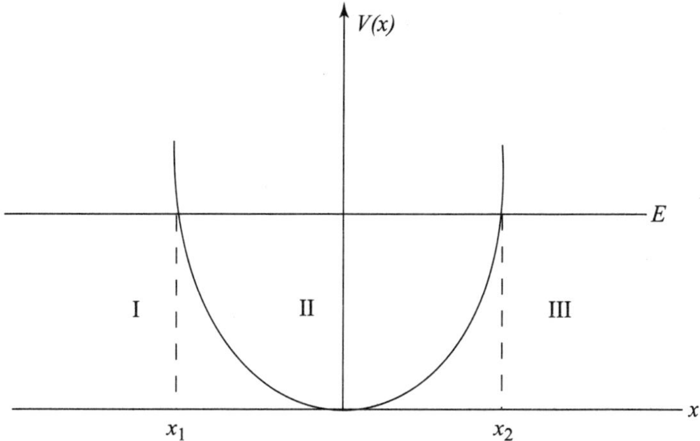

Fig. 6.1 *Potential energy curve depicting turning points*

Let us apply these formulae to the particular case where the potential energy has the form depicted in Fig. 6.1 which also shows the total energy E and the two turning points x_1 and x_2. On the left side of the turning point x_1, i.e. in the region I, for large negative distance x from the turning point,

$$\lim_{x \to -\infty} \psi_{\mathrm{I}}(x) = \frac{1}{2(-x)^{1/4}} \exp\left[\frac{-2a}{3}(-x)^{3/2}\right], \tag{6.33}$$

and on the right-hand side for large positive x,

$$\psi_{\mathrm{II}}(x) = \frac{1}{x^{1/4}} \sin\left(\frac{2a}{3}x^{3/2} + \frac{\pi}{4}\right). \tag{6.34}$$

Comparing with (6.25) and (6.26) we get, using (6.29),

$$\psi_{\mathrm{I}}(x) = \frac{B}{\sqrt{-p}} \exp\left[-\int_{x_1}^{x} dx \frac{p}{\hbar}\right],$$

$$\psi_{\mathrm{II}}(x) = \frac{2A}{\sqrt{+p}} \sin\left(\int_{x_1}^{x} dx \frac{p}{\hbar} + \frac{\pi}{4}\right). \tag{6.35}$$

We now consider the regions on the left side of the turning point x_2, i.e., region II, for which for large negative distance x from the turning point,

$$\psi_{\mathrm{II}}(x) = \frac{1}{x^{1/4}} \sin\left(\frac{2a}{3}x^{3/2} - \frac{\pi}{4}\right). \tag{6.36}$$

Similarly,

$$\psi_{\mathrm{III}}(x) = \frac{1}{2(-x)^{1/4}} \exp\left[\frac{2a}{3}(-x)^{3/2}\right]. \tag{6.37}$$

Comparing these with (6.25) and (6.26) gives

$$\psi_{\rm II}(x) = \frac{2A}{\sqrt{p}} \sin\left(\int_{x_2}^{x} \frac{p}{\hbar} dx - \frac{\pi}{4}\right),$$

$$\psi_{\rm III}(x) = \frac{B}{\sqrt{-p}} \exp\left[-\int_{x_2}^{x} dx \frac{p}{\hbar}\right]. \tag{6.38}$$

Since the $\psi_{\rm II}(x)$ obtained in (6.35) and (6.38) must be equal at $x = x_1$, we have

$$\sin\frac{\pi}{4} = \sin\left(\int_{x_2}^{x_1} \frac{p\,dx}{\hbar} - \frac{\pi}{4}\right), \tag{6.39}$$

which is satisfied if

$$\oint dxp(x) = \left(n + \frac{1}{2}\right) h, \tag{6.40}$$

where $\oint dxp(x) = 2\int_{x_2}^{x_1} dxp(x)$. It will be noticed that (6.40) without the term $\frac{1}{2}$ on the right-hand side is the Bohr–Sommerfeld quantization rule.

Problems

1. The wave equation

$$\frac{d^2\psi}{dx^2} + \left(-\frac{1}{4} + \frac{A}{x} - \frac{l(l+1)}{x^2}\right)\psi = 0$$

has asymptotic solution

$$\psi \underset{x\to\infty}{\longrightarrow} e^{-\frac{1}{2}x}.$$

Use WKB approximation to obtain first order correction

Hint: $\psi = e^{-\frac{1}{2}x}$ is obtained by neglecting $\frac{A}{x} - \frac{l(l+1)}{x^2}$. Next neglect $\frac{l(l+1)}{x^2}$ and integrate taking into account the turning points.

Bibliography

[1] L.D. Landau and E.M. Lifshitz, *Quantum Mechanics*, 3rd edition, Pergamon Press, Oxford (1977).

[2] L.I. Schiff, *Quantum Mechanics*, 3rd edition, McGraw Hill Book Company, Inc. (1977)

[3] H. Jeffreys, *Proc. London Math. Soc.*, (2) **23**, 428 (1923).

[4] G. Wentzel, *Zeits. f. Physik*, **38**, 518 (1926).

[5] H.A. Kramers, *Zeits. f. Physik*, **39**, 828 (1926).

[6] L. Brillouin, *Comptes Rendus*, **183**, 24 (1926).

7 Solution of Schrödinger Equation

7.1 Free Particle with Definite Momentum

We shall obtain solution of Schrödinger equation for a particle with a definite momentum in space without any constraints. The equation for this case is

$$\nabla^2\psi(\vec{r}) + \frac{2mE}{\hbar^2}\psi(\vec{r}) = 0. \tag{7.1}$$

Putting

$$E = \frac{p^2}{2m} = \frac{k^2\hbar^2}{2m}, \tag{7.2}$$

this equation becomes

$$\nabla^2\psi(\vec{r}) + k^2\psi(\vec{r}) = 0, \tag{7.3}$$

where the normalised solution is

$$\psi(\vec{r}) = \frac{1}{(2\pi)^{3/2}}e^{i\vec{k}\cdot\vec{r}}, \tag{7.4}$$

which is also the eigenfunction of the momentum operator $(-i\hbar\vec{\nabla})$. This is due to the commutability of the Hamiltonian $(-\frac{\hbar^2}{2m}\nabla^2)$ with the momentum operator.

7.2 Free Particle with Definite Angular Momentum

The Schrödinger equation (7.1) when expressed in spherical polar coordinates reads as

$$\frac{\partial^2\psi(\vec{r})}{\partial r^2} + \frac{2}{r}\frac{\partial\psi(\vec{r})}{\partial r} + \frac{1}{r^2}\left[\frac{1}{\sin\theta}\frac{\partial}{\partial\theta}\left(\sin\theta\frac{\partial\psi(\vec{r})}{\partial\theta}\right) + \frac{1}{\sin^2\theta}\frac{\partial^2\psi(\vec{r})}{\partial\phi^2}\right]$$

$$+\frac{2mE}{\hbar^2} = 0, \tag{7.5}$$

which, written in terms of the squared angular momentum operator L^2 takes the form

$$-\left(\frac{\partial^2\psi}{\partial r^2}+\frac{2}{r}\frac{\partial\psi}{\partial r}-\frac{L^2}{r^2}\psi\right)=\frac{2mE}{\hbar^2}. \tag{7.6}$$

This equation is separable in r and (θ,φ), i.e.,

$$\psi(\vec{r})=R(r)Y_{lm}(\theta,\varphi), \tag{7.7}$$

substituting which in (7.6) gives, for $R(r)$,

$$\frac{\partial^2 R}{\partial r^2}+\frac{2}{r}\frac{\partial R}{\partial r}-\frac{l(l+1)}{r^2}R+\frac{2mE}{\hbar^2}R=0 \tag{7.8}$$

since, as we have seen in chapter 3,

$$L^2Y_{lm}(\theta\varphi)=l(l+1)Y_{lm}(\theta\varphi). \tag{7.9}$$

For a state with definite angular momentum, we replace R in (7.8) by $R_l(r)$. The orthonormality condition

$$\int d^3r\psi^*_{k'}(\vec{r})\psi_k(\vec{r})=\delta^{(3)}(\vec{k}-\vec{k}') \tag{7.10}$$

takes the form in spherical polar coordinates,

$$\int_0^\pi\int_0^{2\pi}\sin\theta d\theta d\phi Y^*_{l'm'}(\theta\varphi)Y_{lm}(\theta\varphi)=\delta_{ll'}\delta_{mm'}, \tag{7.11}$$

$$\int drr^2R_{k'l'}(r)R^{(r)}_{kl}=2\pi\delta(k'-k)\delta_{ll'}. \tag{7.12}$$

Solution of (7.8) which is finite at $r=0$ is obtained in terms of Bessel function $J_{l+\frac{1}{2}}(kr)$:

$$R_{kl}(r)=\sqrt{\frac{2\pi k}{r}}J_{l+\frac{1}{2}}(kr)=2kj_l(kr), \tag{7.13}$$

where

$$j_l(kr)=\sqrt{\frac{\pi}{2kr}}J_{l+\frac{1}{2}}(kr) \tag{7.14}$$

is known as the spherical Bessel function.

7.3 Particle on a Line Segment

We consider the motion of a particle constrained to move on the segment $0 \leq x \leq a$ of a straight line. The Schrödinger equation for this case is

$$\frac{\partial^2 \psi(x)}{\partial x^2} + \frac{2mE}{\hbar^2}\psi(x) = 0. \tag{7.15}$$

The normalised solutions of this equation with boundary conditions

$$\psi(0) = \psi(a) = 0 \tag{7.16}$$

which express the fact that the particle is constrained to move on the line $0 \leq x \leq a$, are

$$\psi_n(x) = \sqrt{\frac{2}{a}}\,\sin\frac{n\pi}{a}x,\, n = 1, 2, \ldots, \tag{7.17}$$

$$E_n = \frac{n^2\pi^2\hbar^2}{2ma^2}. \tag{7.18}$$

The matrix elements of the Hamiltonian operator $H = -\frac{\hbar^2}{2m}\frac{\partial^2}{\partial x^2}$ work out to be

$$H_{nm} = \int_p^a dx\psi_n^*(x)H\psi_m(x) = E_n\delta_{mm} \tag{7.19}$$

and those of momentum coordinate are found to be

$$\begin{aligned} p_{nm} &= -i\hbar\int_0^a dx\psi_n^*(x)\frac{\partial\psi_m(x)}{\partial x} \\ &= \begin{cases} \frac{4nm\hbar}{ia(n^2-m^2)} & \text{for } (n-m) \text{ and } (n+m) \text{ odd} \\ 0 & \text{for } (n-m) \text{ and } (n+m)\text{even} \\ 0 & \text{for } n=m \end{cases} \end{aligned} \tag{7.20}$$

$$\begin{aligned} x_{nm} &= -\int_0^\infty dx\psi_n^* x\psi_m \\ &= \begin{cases} -\frac{8anm}{(n^2-m^2)^2\pi^2} & \text{for } (n-m) \text{ and } (n+m) \text{ odd} \\ 0 & \text{for } (n-m) \text{ and } (n+m)\text{even} \\ \frac{a}{2} & \text{for } n=m. \end{cases} \end{aligned} \tag{7.21}$$

It is instructive to display the matrices for the above three operators for n and m ranging from 1 to 4:

$$H = \frac{\pi^2\hbar^2}{2ma^2}\begin{pmatrix} 1 & & & \\ & 4 & & 0 \\ 0 & & 9 & \\ & & & 16 \end{pmatrix} \tag{7.22}$$

$$p = \frac{4\hbar}{ia}\begin{pmatrix} 0 & \frac{-2}{3} & 0 & -\frac{4}{15} \\ \frac{2}{3} & 0 & -\frac{6}{5} & 0 \\ 0 & \frac{6}{5} & 0 & -\frac{12}{7} \\ \frac{4}{15} & 0 & \frac{12}{7} & 0 \end{pmatrix} \tag{7.23}$$

$$x = \frac{8a}{\pi^2}\begin{pmatrix} \frac{\pi^2}{16} & \frac{-2}{9} & 0 & -\frac{4}{25} \\ -\frac{2}{9} & \frac{\pi^2}{16} & -\frac{6}{25} & 0 \\ 0 & -\frac{6}{25} & \frac{\pi^2}{16} & -\frac{12}{49} \\ -\frac{4}{25} & 0 & -\frac{12}{49} & \frac{\pi^2}{16} \end{pmatrix} \tag{7.24}$$

It will be noticed that neither p nor x are diagonal in the representation in which H is diagonal since H and p, H and x as well as x and p do not commute.

$$[H, p] \neq 0, [H, x] \neq 0, [x, p] = i\hbar, \tag{7.25}$$

which show that neither p nor x as operators are conserved; there is no space translation invariance of the Hamiltonian.

7.4 Particle on a Circle

Another simple problem is that of a point particle of mass m constrained to move on a circle of radius a for which the Hamiltonian is

$$H = \frac{p_\theta^2}{2a^2m}, \tag{7.26}$$

where

$$p_\theta = -i\hbar\frac{\partial}{\partial\theta}, \tag{7.27}$$

θ being the angle that gives the position of the particle. The Schrödinger equation for this motion is

$$\frac{\partial^2\psi}{\partial\theta^2} + \frac{2ma^2}{\hbar^2}E\psi = 0. \tag{7.28}$$

Its normalised solution is easily seen to be

$$\psi^{(\pm n)}(\theta) = \frac{1}{\sqrt{2\pi}} e^{\pm in\theta}. \tag{7.29}$$

Single-valuedness of the wave function, i.e.

$$\psi(0) = \psi(2\pi) \tag{7.30}$$

requires that n should take on values 0,1,2,.... The energy eigenvalues are

$$E^{(\pm n)} = \frac{n^2\hbar^2}{2ma^2}. \tag{7.31}$$

It will be seen that the two eigenfunctions $e^{in\theta}/\sqrt{2\pi}$ and $e^{-in\theta}/\sqrt{2\pi}$ have the same eigenvalue, i.e., the energy levels are two-fold degenerate. The ground state with $n = 0$ is, however, non-degenerate.

7.5 Particle Confined to a Circular Plane

The Schrödinger equation for a particle whose motion is confined to a circular plane of finite radius 'a' in two-dimensional polar coordinates (r, φ), is

$$\frac{\partial^2\psi}{\partial r^2} + \frac{1}{r}\frac{\partial\psi}{\partial r} + \frac{1}{r}\frac{\partial^2\psi}{\partial\varphi^2} + \frac{2mE}{\hbar^2}\psi = 0. \tag{7.32}$$

Its separable solutions are of the form

$$\psi(r, \varphi) = \phi_l(r)e^{\pm il\varphi}, \tag{7.33}$$

where $l = 0, \pm1.\pm2\ldots$ on account of single-valuedness $\psi(r, \varphi) = \psi(r, \varphi + 2\pi)$. Substitution of (7.33) in (7.32) gives

$$\frac{\partial^2\phi_l}{\partial r^2} + \frac{1}{r}\frac{\partial\phi_l}{\partial r} + \left(\frac{2mE}{\hbar^2} - \frac{l^2}{r^2}\right)\phi_l = 0 \tag{7.34}$$

with boundary condition

$$\psi(r = a) = 0. \tag{7.35}$$

In terms of the variable

$$\rho = \sqrt{\frac{2mE}{\hbar^2}}\, r = \alpha r, \tag{7.36}$$

this equation becomes

$$\frac{d^2\phi_l}{d\rho^2} + \frac{1}{\rho}\frac{d\phi_l}{d\rho} + \left(1 - \frac{l^2}{\rho^2}\right)\phi_l = 0, \tag{7.37}$$

which is Bessel equation with the solution which is finite at the origin:

$$\phi_l(\rho) = N_l J_l(\rho), \tag{7.38}$$

where N_l is a normalisation constant.
The energy eigenvalues E_{nl} are obtained from the boundary condition

$$J_l(a\alpha_{nl}) = 0. \tag{7.39}$$

The values of α_{nl} for which this is satisfied can only be found graphically from a plot of $J_l(\rho)$ against ρ.

7.6 Particle Confined Inside a Sphere

The Schrödinger equation for a particle constrained to move inside a spherical cavity of radius 'a' is

$$\frac{\partial^2\psi}{\partial\varphi^2} + \frac{2}{r}\frac{\partial\psi}{\partial r} - \frac{L^2\psi}{r^2} + \frac{2mE}{\hbar^2}\psi = 0, \tag{7.40}$$

where L^2 is the square of the orbital angular momentum vector. Confinement condition is

$$\psi(r = a) = 0. \tag{7.41}$$

Solutions of this equation are of the form

$$\psi(r, \theta, \varphi) = R_l(r)Y_{lm}(\theta, \varphi). \tag{7.42}$$

Since

$$L^2 Y_{lm}(\theta\varphi) = l(l+1)Y_{lm}(\theta\varphi), \tag{7.43}$$

the equation for the radial function $R_l(r)$ as obtained from (7.40) is

$$\frac{d^2R_l}{dr^2} + \frac{2}{r}\frac{dR_l}{dr} - \frac{l(l+1)}{r^2}R_l + \alpha^2 = 0, \tag{7.44}$$

where

$$\alpha^2 = \frac{2mE}{\hbar^2}. \tag{7.45}$$

In terms of $\rho = \alpha r$, this equation has the form

$$\frac{d^2 R_l}{d\rho} + \frac{2}{\rho}\frac{dR_l}{d\rho} + \left[1 - \frac{l(l+1)}{\rho^2}\right] R_l = 0. \tag{7.46}$$

The solutions of this equation which are finite at $\rho = 0$ are the spherical Bessel functions $j_l(\rho)$ defined in terms of cylindrical essel functions of half-odd integer order by

$$j_l(\rho) = \left(\frac{\pi}{2\rho}\right)^{1/2} J_{l+\frac{1}{2}}(\rho). \tag{7.47}$$

The energy eigenvalues are obtained from the confinement condition (7.41). Since for $l = 0$

$$j_0(\rho) = \frac{\sin\rho}{\rho}, \tag{7.48}$$

the confinement condition

$$\frac{\sin a\alpha}{a\alpha} = 0 \tag{7.49}$$

is satisfied by

$$\alpha_n = \frac{n\pi}{a} \qquad n = 1, 2, \ldots \tag{7.50}$$

so that the energy eigenvalues are

$$E_n = \frac{n^2\hbar^2\pi^2}{2ma^2}. \tag{7.51}$$

It will be noticed that this is identical to the energy eigenvalues for a particle confined to a line segment. The reason is that the particle inside a sphere with zero angular momentum moves along a straight line.*

For $l = 1$, we have

$$j_1(\rho) = \frac{\sin\rho}{\rho^2} - \frac{\cos\rho}{\rho}. \tag{7.52}$$

The confinement condition $j_1(a\alpha) = 0$ gives

$$\tan a\alpha = a\alpha, \tag{7.53}$$

which can only be solved graphically.

---

*Since the length of this line is the diameter of the sphere which is $2a$, its energy eigenvalues corresponding to n is identical to that of the line segment with $\frac{1}{2}n$ for even n. The eigenfunctions are, however, different.

The eigenvalues of $a\alpha$ are the points where the curves

$$y = a\alpha \quad \text{and} \quad y = \tan a\alpha$$

meet. The energies are constrained by the inequality

$$\frac{n^2\pi^2\hbar^2}{2ma^2} \leq E_n \leq \left(\frac{2n+1}{2}\right)^2 \frac{\hbar^2}{2ma^2}. \tag{7.54}$$

7.7 Simple Harmonic Oscillator (Simple Pendulum)

A simple pendulum consists of a particle of mass m hung from a point of rigid support by a massless string of length a and executing oscillations of small amplitudes in a gravitational field. Its Hamiltonian can be written as

$$H = \frac{1}{2}ma^2\dot{\theta}^2 + \frac{1}{2}mga\theta^2, \tag{7.55}$$

where θ is the angle that the string makes with the vertical. In terms of $x = a\theta$, the above Hamiltonian becomes

$$H = \frac{p^2}{2m} + \frac{1}{2}m\omega^2x^2, \tag{7.56}$$

where $\omega^2 = mg/a$ and $p = m\dot{x}$. This Hamiltonian represents a host of physical problems like vibrations of atoms in a diatomic molecule, nuclei in a lattice, oscillations of a finite size nucleus, etc. From the classical Hamilton's equations of motion

$$\dot{x} = \frac{\partial H}{\partial p}, \qquad \dot{p} = -\frac{\partial H}{\partial x}, \tag{7.57}$$

one obtains

$$\ddot{x} + \omega^2 x = 0, \tag{7.58}$$

which shows that the force on the particle is proportional to displacement with a negative sign. Its solution is

$$x = \text{Re}\ ae^{i\omega t}, \tag{7.59}$$

where Re stands for the real part and a is the amplitude of oscillations. In quantum mechanics, the Schrödinger equation of motion becomes

$$\frac{d^2\psi}{dx^2} + \frac{2m}{\hbar^2}\left(E - \frac{1}{2}m\omega^2x^2\right)\psi = 0. \tag{7.60}$$

In terms of dimensionless variable

$$\xi = \left(\frac{m\omega}{\hbar}\right)^{1/2} x, \tag{7.61}$$

(7.60) reads

$$\frac{d^2\psi}{d\xi^2} + \left(\frac{2E}{\hbar\omega} - \xi^2\right)\psi = 0. \tag{7.62}$$

For large ξ, one can neglect $\frac{2E\psi}{\hbar\omega}$ term, whence the above equation ıakes the form

$$\frac{d^2\Psi(\xi)}{d\xi^2} = \xi^2\Psi(\xi), \Psi(\xi) = \lim_{\xi\to\infty} \psi(\xi), \tag{7.63}$$

where the solution which does not blow up at $\xi = \pm\infty$ is

$$\Psi(\xi) = e^{-\frac{1}{2}\xi^2}. \tag{7.64}$$

Therefore the solution of (7.62) will have the form

$$\psi(\xi) = \phi(\xi)e^{-\frac{1}{2}\xi^2}. \tag{7.65}$$

Substituting this in (7.62) gives

$$\frac{d^2\phi}{d\xi^2} - 2\xi\frac{d\phi}{d\xi} + \left(\frac{2E}{\hbar\omega} - 1\right)\phi = 0. \tag{7.66}$$

A polynomial solution of this equation is required to ensure that $\phi(\xi)$ remains finite at $\xi = \infty$. Such solutions exist only for

$$\left(\frac{2E_n}{\hbar\omega} - 1\right) = 2n, n = 0, 1, 2, \dots , \tag{7.67}$$

which is

$$\phi(\xi) = N_n H_n(\xi) = (-)^n N_n e^{\xi^2} \frac{d^n}{d\xi^n} e^{-\xi^2}, \tag{7.68}$$

where $H_n(\xi)$ are Hermite polynomials and N_n are normalisation constants. Thus the energy eigenvalues and the normalised eigenfunctions are

$$E_n = \left(n + \frac{1}{2}\right)\hbar\omega, \tag{7.69}$$

$$\psi_n(\xi) = \left(\sqrt{\pi}\, n!\, ^n\right)^{-\frac{1}{2}} e^{-\frac{1}{2}\xi^2} H_n(\xi). \tag{7.70}$$

The ground state energy $E_0 = \frac{1}{2}\hbar\omega$ is called zero-point energy. The energy eigenvalue formula (7.69) can be interpreted as the nth state having n quanta of energy, each quantum possessing an energy of $\hbar\omega$.

It may be useful to write down expressions for first few Hermite polynomials:

$$H_0(\xi) = 1, \quad H_1(\xi) = 2\xi, \quad H_2(\xi) = 4\xi^2 - 2, \quad H_3(\xi) = 8\xi^3 - 12\xi. \tag{7.71}$$

The recursion relations for these polynomials are

$$\xi H_n(\xi) = nH_{n-1}(\xi) + \frac{1}{2}H_{n+1}(\xi), \tag{7.72}$$

$$\frac{dH_n}{d\xi} = 2nH_{n-1}(\xi). \tag{7.73}$$

With the help of these recursion formulae we obtain

$$\frac{d\psi_n}{d\xi} = \sqrt{\frac{n}{2}}\,\psi_{n-1} - \sqrt{\frac{n+1}{2}}\,\psi_{n+1}, \tag{7.74}$$

$$\xi\psi_n = \sqrt{\frac{n}{2}}\,\psi_{n-1} + \sqrt{\frac{n+1}{2}}\,\psi_{n+1}, \tag{7.75}$$

so that

$$\frac{1}{\sqrt{2}}\left(\xi + \frac{d}{d\xi}\right)\psi_n = \sqrt{n}\,\psi_{n-1}, \tag{7.76}$$

$$\frac{1}{\sqrt{2}}\left(\xi - \frac{d}{d\xi}\right)\psi_n = \sqrt{n+1}\,\psi_{n+1}, \tag{7.77}$$

which can be put in the form

$$a\psi_n = \sqrt{n}\,\psi_{n-1} \quad a^\dagger\psi_n = \sqrt{n+1}\,\psi_{n+1}, \tag{7.78}$$

where,

$$a = \frac{1}{\sqrt{2}}\left(\xi + \frac{d}{d\xi}\right) = \sqrt{\frac{1}{2m\hbar\omega}}\,(m\omega x + ip), \tag{7.79}$$

$$a^\dagger = \frac{1}{\sqrt{2}}\left(\xi - \frac{d}{d\xi}\right) = \sqrt{\frac{1}{2m\hbar\omega}}\,(m\omega x - ip). \tag{7.80}$$

It will be seen that the operator a reduces the energy by one quantum on operating on a state with n quanta while the operator $a^\dagger$ increases the energy by one quantum. These are therefore called annihilation and creation operators. The Hamiltonian can be written in terms of these operators:

$$H = \frac{1}{2}\hbar\omega(a^\dagger a + aa^\dagger). \tag{7.81}$$

One can also verify that the $[x, p] = i\hbar$ commutation relation can be expressed as

$$[a, a^\dagger] = 1. \tag{7.82}$$

It may be seen from (7.74) and (7.75) that the expectation values of x and p for any state $|n\rangle$ vanishes; they have non-zero matrix elements only between neighbouring states, i.e. states differing in n by one unit.

$$\langle n|x|n-1\rangle = i\sqrt{\frac{n\hbar}{2m\omega}}, \quad \langle n|p|n-1\rangle = -im\omega\langle n|x|n-1\rangle. \tag{7.83}$$

However, for a particular superposition of states,

$$|\alpha\rangle = e^{-\frac{1}{2}\alpha^2}\sum_{n=0}^{\infty}\frac{\alpha^n}{\sqrt{n!}}|n\rangle, \tag{7.84}$$

$$\langle\alpha|x|\alpha\rangle \neq 0. \tag{7.85}$$

It can be seen by actual calculation that this state is an eigenstate of the annihilation operator

$$a|\alpha\rangle = \alpha|\alpha\rangle \tag{7.86}$$

and

$$\langle\beta|\alpha\rangle = e^{-\frac{1}{2}(\alpha^2+\beta^2)+\alpha\beta} > 0 \tag{7.87}$$

so that

$$\langle\alpha|\alpha\rangle = 1. \tag{7.88}$$

The operator $a^\dagger a$ has eigenvalue n for the state $|n\rangle$

$$a^\dagger a|n\rangle = n|n\rangle \tag{7.89}$$

and is therefore the number operator $N = a^\dagger a$.

The expectation value of this operator for $|\alpha\rangle$ works out tossssss

$$\bar{n} = \langle\alpha|N|\alpha\rangle = \alpha^2. \tag{7.90}$$

Other relevant expectation values work out to

$$\langle\alpha|H|\alpha\rangle = \left(\alpha^2 + \frac{1}{2}\right)\hbar\omega = \left(\bar{n} + \frac{1}{2}\right)\hbar\omega, \tag{7.91}$$

$$\bar{p} = \langle\alpha|p|\alpha\rangle = 0, \quad \bar{x} = \langle\alpha|x|\alpha\rangle = \sqrt{\frac{2\hbar}{m\omega}}\,\alpha, \tag{7.92}$$

$$\bar{p^2} = \langle\alpha|p^2|\alpha\rangle = \frac{m\hbar\omega}{2}, \tag{7.93}$$

$$\bar{x^2} = \langle\alpha|x^2|\alpha\rangle = \frac{\hbar}{2m\omega}(4\alpha^2 + 1). \tag{7.94}$$

From these expectation values we find that

$$\langle\alpha|\ddot{x}|\alpha\rangle + \omega^2\langle\alpha|x|\alpha\rangle = 0. \tag{7.95}$$

The state $|\alpha\rangle$, therefore, can be interpreted as a classical state. Further, it has the minimum uncertainty of $\frac{1}{2}h$ which can be verified as follows:

From (7.92), (7.93) and (7.94),

$$\delta x = \sqrt{(\bar{x^2} - \bar{x}^2)} = \sqrt{\frac{\hbar}{2m\omega}}, \tag{7.96}$$

$$\delta p = \sqrt{(\bar{p^2} - \bar{p}^2)} = \sqrt{\frac{m\hbar\omega}{2}}, \tag{7.97}$$

from which one gets

$$\delta x \delta p = \frac{\hbar}{2}. \tag{7.98}$$

The probability of this state having n quanta is given by

$$P(n) = |\langle n|\alpha\rangle|^2 = \frac{\alpha^{2n}}{n!}\,e^{-\alpha^2} = \frac{(\bar{n})^{2n}e^{-\bar{n}}}{n!}, \tag{7.99}$$

which is a Poisson distribution.

It may be worth mentioning here that this state occurs in quantum optics with n being the number of photons for which the degree of coherence has maximum value of unity. It is on account of this that these states are called coherent states. The orthonormality and completeness for these states can be written as

$$\langle\beta|\alpha\rangle = \exp\left[-\frac{1}{2}(|\alpha|^2 + |\beta|^2) + \alpha\beta^*\right],$$

$$\int d^2\alpha|\alpha\rangle\langle\alpha| = 1, \tag{7.100}$$

for which reason they are said to be an over-complete set.

For oscillators in higher dimensions the Hamiltonian is

$$H = \frac{p_i p_i}{2m} + \frac{1}{2}\omega^2 x_i x_i, i = 1, 2, 3, \ldots, \tag{7.101}$$

which is a sum of one-dimensional oscillators. It is easy to see that the eigenvalues are the sum of the eigenvalues and eigenfunction are the product of eigenfunction of one-dimensional oscillators.

7.8 Plane Pendulum

A plane pendulum consists of a particle attached to a rigid string of length 'a' such that it is constrained to move on a vertical circular path and is under the influence of uniform gravitational field. The motion of the particle is described in terms of the angle θ between the downward vertical and the radius vector of the circle. Its Hamiltonian is

$$H = \frac{1}{2}ma^2\dot{\theta}^2 + mga(1 - \cos\theta). \tag{7.102}$$

Classically,

$$E = 0 \quad \text{for} \quad \theta = 0, \tag{7.103}$$

which corresponds to the particle being at the lowest point of the circle.
For $0 < E < 2mga$, the motion is oscillatory with period τ.

$$y = \sin\frac{\theta}{2} = \left(\frac{E}{2mga}\right)^{1/2} \text{sn}\left(\frac{g}{a}\right)^{1/2} t, \tag{7.104}$$

where the elliptic function sn $\left(\frac{g}{a}\right)^{1/2}$ is defined by

$$\text{sn } u = z \tag{7.105}$$

with

$$u = \int_0^z \frac{dz}{\sqrt{(1 - z^2)(1 - k^2 z^2)}}, \quad \tau = 4\int_0^1 \frac{dz}{\sqrt{(1 - z^2)(1 - k^2 z^2)}} \tag{7.106}$$

and $k^2 = E/2mga$. The period $\tau \approx 2\pi\sqrt{\frac{a}{g}}(1 + \frac{k^2}{4})$ depends on the amplitude $2k$.

For $E = 2mga$, the particle swings up to the position $\theta = \pi$ and remains there forever. This is known as a sticking solution.

$$y = \tanh\left(\frac{g}{a}\right)^{1/2} t, \tag{7.107}$$

from which one sees that for $t \to \infty$, $y \to 1$, i.e., $\theta = \pi$.

In quantum mechanics, the motion is described by the Schrödinger equation

$$\left[-\frac{\hbar^2}{2ma^2}\frac{d^2}{d\theta^2} + mga(1-\cos\theta)\right]\psi = E\psi. \tag{7.108}$$

Putting $\theta = 2\phi$, this equation reduces to the Mathieu equation

$$\frac{d^2\psi}{d\phi^2} + (\eta + \gamma\cos 2\phi)\psi = 0, \tag{7.109}$$

where

$$\eta = \frac{8ma^2(E - mga)}{\hbar^2}, \quad \gamma = \frac{8m^2ga^3}{\hbar^2}. \tag{7.110}$$

The boundary condition for θ is

$$\psi(\theta) = \psi(\theta + 2\pi),$$

which for ϕ is

$$\psi(\phi) = \psi(\phi + \pi). \tag{7.111}$$

The solutions of (7.108) satisfying the boundary conditions are Mathieu functions

$$\psi_n(\gamma,\phi) = ce_{2n}(\gamma\phi) = \sum_{s=0}^{\infty} A_{2s}^{2n}(\gamma)\cos 2s\phi,$$

$$\psi_n(\gamma,\phi) = se_{2n+2}(\gamma,\phi) = \sum_{s=0}^{\infty} B_{2s+2}^{2n+2}(\gamma)\sin(2s+2)\phi. \tag{7.112}$$

For $\gamma = 0$, these reduce to $\cos 2n\phi$ and $\sin(2n+2)\phi$. The values of η corresponding to these eigenfunctions are plotted in Fig. 7.1 as function of γ as dotted and solid lines respectively. It will be seen that these curves intersect the η axis ($\gamma = 0$) at the points $\eta = 4n^2$ and have contacts of order $(2n-1)$ at these points.

The γ versus η curve can be divided into three regions by the straight lines $\eta = -\gamma$ and $\eta = \gamma$. For $\eta < -\gamma$, there are no acceptable physical solutions. In this region $\eta + \gamma = \frac{8ma^2}{\hbar^2}E < 0$. This is also true for the classical case.

The solutions of Schrödinger equation that do not satisfy the boundary conditions (7.111) are

$$\psi = ce_{2n+1}(\gamma,\phi) \quad \text{and} \quad se_{2n+1}(\gamma,\phi). \tag{7.113}$$

They change sign under rotation of 2π, i.e.,

$$\psi(\theta) = -\psi(\theta + 2\pi) \tag{7.114}$$

like the eigenfunction of spin-$\frac{1}{2}$ particles.

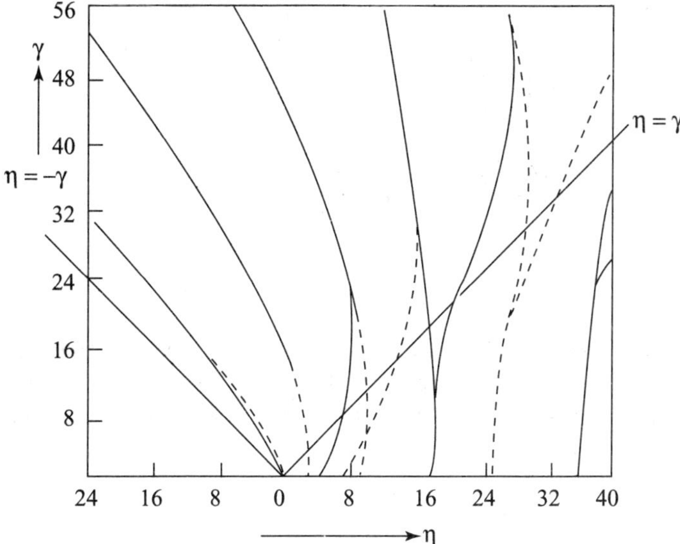

Fig. 7.1 *The* (γ, η) *plot for the eigenfunctions*

Finally, we consider two limiting cases of the plane pendulum problem:

1. Limit $\gamma \to \infty$.

   Both the dotted and the undotted curves become parallel to the line $\eta = -\gamma$ and intersect lines parallel to the η axis at the points

   $$\eta + \gamma = (2\gamma)^{1/2}(2n + 1). \tag{7.115}$$

   Therefore the energy spectrum for $\gamma \to \infty$ is

   $$E_n = (\gamma + n)\frac{\hbar^2}{8ma^2} = \left(n + \frac{1}{2}\right)\left(\frac{g}{a}\right)^{1/2}\hbar = \left(n + \frac{1}{2}\right)\hbar\omega, \tag{7.116}$$

   which is the energy spectrum of the simple pendulum, i.e. the harmonic oscillator discussed in the previous section. It can also be shown that in this limit

   $$ce_{2n}(\gamma, \phi) = se_{2n+2}(\gamma, \phi) = W(\xi), \tag{7.117}$$

   where $W(\xi)$ satisfy the oscillator equation

   $$\frac{d^2W(\xi)}{d\xi^2} + \left(\lambda - \xi^2\right)W(\xi) = 0 \tag{7.118}$$

with

$$\lambda = 2E \left(\frac{a}{g}\right)^{1/2} \hbar, \tag{7.119}$$

$$\xi = \left(\frac{16m^2 g a^3}{\hbar^2}\right)^{1/4} \phi. \tag{7.120}$$

2. Limit $\gamma = 0$.

   In this limit $g = 0$, which corresponds to motion of a free particle constrained to move on a circular path; this has been discussed in section 7.4. The energy spectrum obtained from the Schrödinger equation is

$$E_n^{(\pm n)} = \frac{n^2 \hbar^2}{2ma^2} \qquad n = 0, 1, 2, \ldots . \tag{7.121}$$

7.9 Particle in a Cylindrical Well

In this section we consider the motion of a particle inside a cylindrical well of radius a and a depth v. The Schrödinger equation for such a particle inside the well with negative energy $E = -\epsilon$ is

$$\frac{\partial^2 \psi}{\partial r^2} + \frac{1}{r}\frac{\partial \psi}{\partial r} + \frac{1}{r^2}\frac{\partial^2 \psi}{\partial \varphi^2} + \frac{2m}{\hbar^2}(v - \epsilon)\psi = 0. \tag{7.122}$$

As in section 7.5, the solutions are of the form

$$\psi(r, \varphi) = \phi_l(r) e^{\pm i l \varphi}. \tag{7.123}$$

Substitution of (7.123) in (7.122) gives

$$\frac{\partial^2 \phi_l}{\partial r^2} + \frac{1}{r}\frac{\partial \phi_l}{\partial r} + \left[\frac{2m}{\hbar^2}(v - \epsilon) - \frac{l^2}{r^2}\right]\phi_l = 0. \tag{7.124}$$

In terms of the variable

$$\rho = \sqrt{\frac{2m}{\hbar^2}(v - \epsilon)}\, r = \alpha r, \tag{7.125}$$

(7.124) takes the form

$$\frac{\partial^2 \phi_l}{\partial \rho^2} + \frac{1}{\rho}\frac{\partial \phi_l}{\partial \rho} + \left(1 - \frac{l^2}{\rho^2}\right)\phi_l = 0. \tag{7.126}$$

For the motion of the particle with negative energy outside the well for which $v = 0$, the variable ρ becomes

$$\rho = \sqrt{\frac{2m\epsilon}{\hbar^2}}\, ir = i\beta r \tag{7.127}$$

in terms of which the equation becomes

$$\frac{\partial^2 \phi_l^{(0)}}{\partial \rho^2} + \frac{1}{\rho}\frac{\partial \phi_l^{(0)}}{\partial \rho} - \left(1 + \frac{l^2}{\rho^2}\right)\phi_l^{(0)} = 0. \tag{7.128}$$

Solutions for (7.126) which are finite at the origin are

$$\phi_l(\rho) = J_l(\rho), \tag{7.129}$$

where $J_l(\rho)$ are the cylindrical Bessel functions. Solutions of (7.128) which vanish at infinity are

$$\phi_l^{(0)}(\rho) = I_l(\rho) + iK_l(\rho), \tag{7.130}$$

where $I_l(\rho)$ and $K_l(\rho)$ are the modified Bessel and Neuman functions respectively. The wavefunctions $\phi_l(\rho)$ and $\phi_l^{(0)}(\rho)$ as well as their derivatives must match at $\rho = a\alpha$. This would yield the energy eigenvalues.

7.10 Particle in a Spherical Well

Spherical well is a space of finite radius inside which the potential is constant and negative, and outside the well it is zero, i.e.,

$$V = \begin{cases} -v & r < a \\ 0 & r > a \end{cases} \tag{7.131}$$

The Schrödinger equation for the particle is

$$-\frac{\partial^2 \psi}{\partial r^2} + \frac{2}{r}\frac{\partial \psi}{\partial r} - \frac{L^2}{r^2}\psi + \frac{2m}{\hbar^2}(E - V)\psi = 0, \tag{7.132}$$

the solution of which are separable in radial and angular coordinates as in the previous section:

$$\psi(r, \theta, \varphi) = R_l(r) Y_{lm}(\theta, \varphi) \tag{7.133}$$

so that the radial Schrödinger equation for $r < a$ becomes

$$\frac{d^2 R_l(r)}{dr^2} + \frac{2}{r}\frac{dR_l(r)}{dr} + \left[\alpha^2 - \frac{l(l+1)}{r^2}\right] R_l(r) = 0, \tag{7.134}$$

where

$$\alpha = \frac{2m}{\hbar^2}(v - \epsilon) \geq 0. \tag{7.135}$$

In terms of the variable $\rho = \alpha r$, the above equation takes the form

$$\frac{d^2 R_l(\rho)}{d\rho^2} + \frac{2}{\rho}\frac{dR_l(\rho)}{d\rho} + \left[1 - \frac{l(l+1)}{\rho^2}\right] R_l(\rho) = 0. \tag{7.136}$$

Its solutions are the spherical Bessel functions $j_l(\rho)$ and $n_l(\rho)$:

$$j_l(\rho) = \sqrt{\frac{\pi}{2\rho}} J_{l+\frac{1}{2}}(\rho), n_l(\rho) = \sqrt{\frac{\pi}{2\rho}}\, N_{l+\frac{1}{2}}(\rho), \tag{7.137}$$

where $J_l(\rho)$ and $N_l(\rho)$ are cylindrical Bessel and Neuman function respectively. Since we need to have finite $R_l(\rho)$ at $\rho = 0$, only $j_l(\rho)$ is acceptable. For $r > a$, $V = 0$, the radial Schrödinger equation has the same from as (7.136) but with

$$\rho = i\beta r, \quad \beta^2 = -\frac{2m\epsilon}{\hbar^2}. \tag{7.138}$$

Because of this, the solutions that vanish at infinity are

$$R_l(\rho) = h_l^{(1)}(\rho) = j_l(\rho) + i\eta_l(\rho). \tag{7.139}$$

These we call as exterior solutions, and (7.137) the interior solutions. Energy eigenvalues are obtained by matching the exterior and interior wave functions as well as their derivatives at $r = a$.

For $l = 0$, the matching gives

$$\xi \cot \xi = -\eta,$$

$$\xi^2 + \eta^2 = \frac{2mva^2}{h^2}, \tag{7.140}$$

where $\xi = a\alpha \;\; \eta = a\beta$.

The energy eigenvalues are the points of intersection of the curves representing the two equation (7.140). For $l = 1$, the matching gives

$$\frac{\cot \xi}{\xi} - \frac{1}{\xi^2} = \frac{1}{\eta} + \frac{1}{\eta^2}$$

$$\xi^2 + \eta^2 = \frac{2mva^2}{\hbar^2}. \tag{7.141}$$

7.11 Particle in a Coulomb Potential in Two Dimensions

The Schrödinger equation for the motion of a particle moving in an attractive Coulomb potential $V = -e^2/r$ in two space dimensions for $E = -\epsilon$ is

$$\frac{\partial^2 \psi}{\partial r^2} + \frac{1}{r}\frac{\partial \psi}{\partial r} + \frac{1}{r^2}\frac{\partial^2 \psi}{\partial \varphi^2} - \frac{2m}{\hbar^2}\left(\epsilon - \frac{e^2}{r}\right)\psi = 0. \tag{7.142}$$

The boundary condition for bound state solution is $\psi(r = 0)$ is finite and $\psi(r = \infty) = 0$. Here r and φ are the particle's two-dimensional polar coordinates. Putting

$$\psi(r, \varphi) = e^{\pm il\varphi}\phi_l(r) \tag{7.143}$$

as in section 9 of this chapter in (7.142) gives,

$$\frac{\partial^2 \phi_l}{\partial r^2} + \frac{1}{r}\frac{\partial \phi_l}{\partial r} - \frac{2m}{\hbar^2}\left(\epsilon - \frac{e^2}{r} + \frac{l^2\hbar^2}{2mr^2}\right)\phi_l = 0. \tag{7.144}$$

In terms of the variable

$$\rho = \left(\frac{8m\epsilon}{\hbar^2}\right)^{1/2} r,$$

(7.144) reads as

$$\frac{d^2\phi_l}{d\rho^2} + \frac{1}{\rho}\frac{d\phi_l}{d\rho} + \left[-\frac{1}{4} - \frac{l^2}{\rho^2} + \frac{\beta}{2\rho}\right]\phi_l = 0, \tag{7.145}$$

where

$$\beta^2 = \frac{2me^4}{\hbar^2\epsilon}. \tag{7.146}$$

For $\rho \to 0$, this equation reduces to

$$\frac{d^2\phi_l}{d\rho^2} + \frac{1}{\rho}\frac{d\phi_l}{d\rho} - \frac{l^2}{\rho^2}\phi_l = 0 \tag{7.147}$$

whose solution goes like

$$\phi_l(\rho) \sim \rho^l \tag{7.148}$$

which is finite at $\rho = 0$. For $\rho \to \infty$, (7.145) reduces to

$$\frac{d^2\phi_l}{d\rho^2} = \frac{\phi_l}{4}, \tag{7.149}$$

whose solution is

$$\phi_l = e^{\pm\rho/2}. \tag{7.150}$$

The solution which satisfies the boundary conditions (b.c.) at infinity is the one with negative sign in the exponent. Thus the wavefunction $\phi_l(\rho)$ which is finite at $\rho = 0$ and polynomial bound* at $\rho = \infty$

$$\phi_l(\rho) = \rho^l e^{-\rho/2}\omega(\rho). \tag{7.151}$$

Substitution of this in (7.145) gives

$$\rho\omega''(\rho) + (2l + 1 - \rho)\omega'(\rho) + \left(\frac{\beta}{2} - \frac{2l+1}{2}\right)\omega(\rho) = 0. \tag{7.152}$$

The solution satisfying the b.c. at $\rho = 0$ is the confluent hypergeometric function F, which can be written as an infinite series in positive integral powers of ρ:

$$\omega(\rho) = F\left(\frac{2l+1}{2} - \frac{\beta}{2}, 2l + 1, \rho\right), \tag{7.153}$$

$$F(a, b, \rho) = 1 + \frac{a\rho}{b} + \frac{1}{2!}\frac{a(a+1)}{b(b+1)}\rho^2 + \cdots \tag{7.154}$$

For satisfying the b.c. at infinity, the series has to terminate which requires

$$\frac{2l+1}{2} - \frac{\beta}{2} = -n_r, \quad n_r = 0, 1, 2, \ldots \tag{7.155}$$

Substituting the values of β from (7.146) gives for energy eigenvalue

$$\epsilon_n = \frac{me^4}{2\left(n - \frac{1}{2}\right)^2 \hbar^2}, \tag{7.156}$$

where

$$n = n_r + l + 1 = 1, 2, \cdots. \tag{7.157}$$

It will be noticed that energy eigenvalues are independent of l, which means there is a degeneracy. From (7.156) we see that for a given value of n there are n different states with l ranging from 0 to $n - 1$ with the same energy. Thus the nth energy eigenstate is n-fold degenerate.

Combining (7.143), (7.151) and (7.153), the eigenfunctions work out to

$$\psi(r, \varphi) = N_{nl}\frac{e^{il\varphi}}{\sqrt{2\pi}}\rho^l e^{-\frac{1}{2}\rho} F(-n + l + 1, 2l + 1, \rho). \tag{7.158}$$

*This terminology is elaborated in chapter 8, section 2.

7.12 Harmonic Oscillator in Two Dimensions

As per discussions in section 7.7, the Hamiltonian of the oscillator in two dimensions can be decomposed into those of two oscillators whose eigenfunctions are functions of their linear momenta, $p_i = i\omega x_i$. In this section we shall obtain the eigenfunctions in terms of the angular momentum. For this, we shall use two-dimensional polar coordinates in which the Schrödinger equation can be written as

$$\frac{\partial^2 \psi}{\partial r^2} + \frac{1}{r}\frac{\partial \psi}{\partial r} + \frac{1}{r^2}\frac{\partial^2 \psi}{\partial \varphi^2} + \frac{2m}{\hbar^2}\left(E - \frac{1}{2}m\omega^2 r^2\right)\psi = 0. \tag{7.159}$$

As in the previous section, we shall write

$$\psi(r, \varphi) = e^{\pm il\varphi}\phi(r) \tag{7.160}$$

where l is the angular momentum quantum number, and substitute it in (7.159) to obtain the radial equation

$$\left[\frac{d^2}{dr^2} + \frac{1}{r}\frac{d}{dr} - \frac{l^2}{r^2} + \frac{2mE}{\hbar^2} - \frac{m^2\omega^2 r^2}{\hbar^2}\right]\phi(r) = 0. \tag{7.161}$$

It is convenient to use the variable $\rho = \frac{m\omega}{\hbar}r^2$ in terms of which this equation takes the form

$$\left(\rho\frac{d^2}{d\rho^2} + \frac{d}{d\rho} - \frac{l^2}{4\rho} + \frac{1}{2}\frac{E}{\hbar\omega} - \frac{\rho}{4}\right)\phi(\rho) = 0. \tag{7.162}$$

Taking into account the boundary conditions

$$\phi(\rho = 0) = \text{finite}, \quad \phi(\rho = \infty) = 0, \tag{7.163}$$

we put

$$\phi(\varphi) = \rho^{l/2}e^{-\frac{1}{2}\rho}\omega(\rho) \tag{7.164}$$

and substitute it in (7.162) to obtain

$$\left[\frac{d^2}{d\rho^2} + (l + 1 - \rho)\frac{d}{d\rho} + \left(\frac{E}{2\hbar\omega} - \frac{l+1}{2}\right)\right]\omega(\rho) = 0, \tag{7.165}$$

which is a confluent hypergeometric equation whose solution is

$$\omega(\rho) = F\left(\frac{l+1}{2} - \frac{E}{2\hbar\omega}, \, l + 1, \rho\right). \tag{7.166}$$

As in the previous section, the series which this function represents has to be terminated to avoid blowing up of the wavefunction at infinity, i.e., we have to have

$$\frac{l+1}{2} - \frac{E}{2\hbar\omega} = -n_r, n_r = 0, 1, 2, \ldots \tag{7.167}$$

so that

$$E = (2n_r + l + 1)\hbar\omega = (n+1)\hbar\omega, \tag{7.168}$$

where $n = 2n_r + l$. The eigenfunctions are

$$\psi = \frac{e^{\pm il\varphi}}{\sqrt{2\pi}} N_{nl}\rho^{l/2} e^{-\frac{1}{2}\rho} F\left(\frac{l-n}{2}, l+1, \rho\right). \tag{7.169}$$

7.13 Electron in a Magnetic Field

The Schrödinger equation for an electron in a magnetic field $\vec{B} = \vec{\nabla} \times \vec{A}$ can be written in terms of $\vec{A}$:

$$E\psi = -\frac{1}{2m}\nabla^2\psi - \frac{ie}{m}\vec{A} \cdot \vec{\nabla}\psi + \frac{e^2 A^2}{2m}\psi, \tag{7.170}$$

where we have taken $\vec{\nabla} \cdot \vec{A} = 0$. First we shall seek solution in rectangular coordinates and later in cylindrical coordinates to obtain wavefunction in terms of angular momentum in a plane perpendicular to the magnetic field.

Rectangular coordinates

A uniform static magnetic field $\vec{B}$ along the z-axis can be represented by

$$A_x = -yB, \qquad A_y = A_z = 0,$$

so that (7.170) takes the form

$$E\psi = -\frac{1}{2m}\nabla^2\psi + \frac{iey B}{m}\frac{\partial\psi}{\partial x} + \frac{e^2 y^2 B^2 \psi}{2m}. \tag{7.171}$$

Putting

$$\psi = e^{i(xp_x + zp_z)}\phi(y) \tag{7.172}$$

in (7.171) gives

$$\frac{\partial^2\phi}{\partial y^2} + \frac{2m}{\hbar^2}\left[\left(E - \frac{p_z^2}{2m}\right) - \frac{1}{2}\omega^2(y - y_0)^2\right]\phi(y) = 0, \tag{7.173}$$

where

$$y_0 = -\frac{p_x}{eB}, \quad \omega = \frac{eB}{m}. \tag{7.174}$$

This equation represents a one-dimensional oscillator in y and a plane wave along z. Therefore its solution is

$$E_{n,p_z} = \left(n + \frac{1}{2}\right)\hbar\omega + \frac{p_z^2}{2m}, \tag{7.175}$$

$$\phi_n(y) = \frac{1}{\pi^{1/4}a^{1/2}\sqrt{2^n n!}} e^{-\frac{(y-y_0)^2}{2a^2}} H_n\left(\frac{y-y_0}{a}\right), \tag{7.176}$$

where $a^2 = \frac{\hbar}{m\omega}$. It will be seen that the energy is the same for all values of p_x which are continuous. Thus there is infinite degeneracy.

Cylindrical coordinates

Here our choice of the vector potential will be

$$\vec{A} = \frac{1}{2}\vec{B} \times \vec{r}. \tag{7.177}$$

Substituting this in the Schrödinger equation (7.170), we get

$$E\psi = -\frac{1}{2m}\nabla^2\psi + \frac{1}{2m}\vec{B}\cdot\vec{L}\psi + \frac{e^2B^2r^2\psi}{8m} \tag{7.178}$$

with $r^2 = x^2 + y^2$ and $\vec{L} = \vec{r} \times \vec{p}$. In cylindrical polar coordinates, this equation becomes for $\vec{B} = (0, 0, B)$,

$$\frac{\partial^2\psi}{\partial r^2} + \frac{1}{r}\frac{\partial\psi}{\partial r} + \frac{1}{r^2}\frac{\partial^2\psi}{\partial\varphi^2} + \frac{\partial^2\psi}{\partial z^2} - \beta^2r^2\psi + 2i\beta\frac{\partial\psi}{\partial\varphi} = -\frac{2mE}{\hbar^2}\psi, \tag{7.179}$$

with $\beta = eB/2\hbar$. Putting

$$\psi(r, \varphi, z) = \frac{e^{il\varphi}}{\sqrt{2\pi}}\frac{e^{ikz}}{\sqrt{2\pi}}\phi(r) \tag{7.180}$$

in (7.178) gives

$$\frac{d^2\phi}{dr^2} + \frac{1}{r}\frac{d\phi}{dr} - \left(\frac{l}{r^2} + 2\beta l + \beta^2r^2 - \frac{2mE}{\hbar^2} + k^2\right)\phi(r) = 0. \tag{7.181}$$

In terms of $\rho = \beta r^2$, this equation reads

$$\rho\frac{d^2\phi}{dr^2} + \frac{d\phi}{d\rho} + \left(\lambda - \frac{l}{2} - \frac{\rho}{4} - \frac{l^2}{4\rho}\right)\phi = 0, \tag{7.182}$$

where

$$\lambda = \frac{2mE - \hbar^2k^2}{4\beta\hbar^2}. \tag{7.183}$$

Solution of (7.181) which satisfies boundary conditions at $\rho = 0$ and $\rho = \infty$ can be written in the form

$$\phi(\rho) = \rho^{l/2}e^{-\frac{1}{2}\rho}\omega(\rho), \tag{7.184}$$

where $\omega(\rho)$ satisfies the differential equation

$$\frac{d^2\omega(\rho)}{d\rho^2} + (l + 1 - \rho)\frac{d\omega(\rho)}{d\rho} + \left(\lambda - l - \frac{1}{2}\right)\omega(\rho) = 0 \tag{7.185}$$

with the solution

$$\omega(\rho) = F\left(l + \frac{1}{2} - \lambda, l + 1, \rho\right). \tag{7.186}$$

For preventing the wavefunction from blowing up at infinity, we have to have

$$l + \frac{1}{2} - \lambda = -n_r, n_r = 0, 1, 2, \ldots . \tag{7.187}$$

Substituting this in (7.183), we get

$$E_{n,k} = \frac{eB}{m}\left(n + \frac{1}{2}\right) + \frac{\hbar^2k^2}{2m}, \tag{7.188}$$

where $n = n_r + l$.

This represents a discrete spectrum for the motion (Landau levels) in the plane perpendicular to the magnetic field and a continuous spectrum for the motion in the direction parallel to it. Collecting terms from (7.180), (7.184) and (7.186), we get

$$\psi(\rho, \varphi, z) = \frac{N_{ne}}{2\pi}e^{ikz+il\varphi}\rho^{l/2}e^{-\rho/2}F(-l - n, l + 1, \rho), \tag{7.189}$$

where N_{ne} is the normalisation constant.

Problems

1. Find the two lowest energy eigenvalue for a particle moving in a potential defined by

$$V = \begin{cases} \infty & \text{for } x = \pm\frac{1}{2}a \\ 0 & \text{for } 0 > x > -\frac{1}{2}a \\ v & \text{for } 0 < x < \frac{1}{2}a. \end{cases}$$

   Hint: For $E < v$, obtain solutions for $0 < x < \frac{1}{2}a$ and $0 > x > -\frac{1}{2}a$ and match the wavefunction and its derivative at the boundary of these regions. This would put a condition which has to be solved graphically. It will be found that this condition cannot be met, which means the solution is not possible—there is no energy level with $E < v$. Proceed in the same manner for $E > v$ for which the condition can be met.

2. For a particle moving in the potential defined by

$$V = \begin{cases} \infty & \text{for } x = 0 \\ -\beta & \text{for } 0 < x < a, \end{cases}$$

   find β for which the lowest energy eigenvalue is $E = -\frac{2\hbar^2}{ma^2}$.
   Hint: Employ the same method as in the previous problem.

3. For a particle moving in a potential defined by

$$V = \begin{cases} \infty & \text{for } x = \pm a \\ 0 & \text{for } b < x < a \text{ and } -a < x < -b \\ v & \text{for } -b < x < b, \end{cases}$$

   with $b < a, a > 0, b > 0$, find the energy eigenvalues and eigenfunctions of the two lowest states.
   Hint: In this case, there are three regions and four boundaries. Take up $E < v$ and $E > v$ separately. There will be symmetric and antisymmetric (with respect to reflection about $x = 0$) solutions in both cases. Obtain the eigenvalues and eigenfunctions for each for both $E < v$ and $E > v$.

4. Show that for a harmonic oscillator the coherent state can be represented by

$$\psi_\alpha(x) = \left(\frac{m\omega}{\pi\hbar}\right)^{1/4} \bar{e}^{(x-\bar{\alpha})^2 \frac{m\omega}{2\hbar}}.$$

Hint: Use (7.70) in (7.84) and make use of

$$e^{-\frac{1}{2}\alpha^2+\sqrt{2}\alpha\xi} = \sum_n \frac{\alpha n H_n(\xi)}{2^{n/2}n!}$$

and (7.92).

Bibliography

[1] E.U. Condon, *Phys. Rev.*, **31**, 891 (1928).

[2] L. Brillouin, *Wave Propagation in Periodic Structures*, Dover, New York, (1953).

[3] L.D. Landau and E.M. Lifshitz, *Quantum Mechanics*, 3rd edition, Pergamon Press, Oxford (1977).

[4] L.I. Schiff, *Quantum Mechanics*, 3rd edition, McGraw Hill Book, Inc., (1977).

[5] H.C. Corben and P. Stehle, *Classical Mechanics*, John Wiley and Sons, New York (1950).

[6] L.D. Landau, *Z. Phys.*, **64**, 629 (1930).

[7] E.H. Kennard, *Z. Phys.*, **14**, 326 (1927).

[8] C.G. Darwin, *Proc. Roy. Soc. London*, **117**, 258 (1928).

[9] L. Page, *Phys. Rev.*, **36**, 444 (1930).

[10] G.E. Uhlenbeck and X. Young, *Phys. Rev.*, **36**, 1721 (1930).

[11] T. Pradhan and A. Khare, *Am. J. Phys.*, **41**, 59 (1973).

8 The Hydrogen Atom

8.1 Two-Body Schrödinger Equation

The Hamiltonian for the hydrogen atom consisting of a proton of mass m_p and an electron of mass m_e interacting with each other through the attractive Coulomb potential is

$$H = -\frac{\hbar^2}{2m_p}\nabla_p^2 - \frac{\hbar^2}{2m_e}\nabla_e^2 - \frac{e^2}{|\vec{r}_p - \vec{r}_e|}. \tag{8.1}$$

This can be expressed in terms of centre of mass and relative coordinates.

$$\vec{R} = \frac{m_p\vec{r}_p + m_e\vec{r}_e}{m_p + m_e} \qquad \vec{r} = \vec{r}_p - \vec{r}_e \tag{8.2}$$

$$H = -\frac{\hbar^2}{2M}\nabla_R^2 - \frac{\hbar^2}{2m}\nabla_r^2 - \frac{e^2}{r}, \tag{8.3}$$

where $M = m_p + m_e$ and $m = \frac{m_p m_e}{m_p + m_e}$ is the reduced mass of the atom. It is clear from (8.3) that the wavefunction is separable in the centre of mass and relative coordinates:

$$\Psi(\vec{r}_p, \vec{r}_e) = \phi(\vec{R})\psi(\vec{r}), \tag{8.4}$$

where

$$-\frac{\hbar^2}{2M}\nabla_R^2\phi(\vec{R}) = \frac{P^2}{2M}\phi(\vec{R}) = E_{CM}\phi(\vec{R}), \tag{8.5}$$

$\vec{P}$ and E_{CM} being the momentum and energy of the centre of mass motion and

$$\left(-\frac{\hbar^2}{2m}\nabla_r^2 - \frac{e^2}{r}\right)\psi(\vec{r}) = E\psi(\vec{r}), \tag{8.6}$$

where E is the energy of the relative motion. We shall be concerned with this equation and its solutions. In spherical polar coordinates (r, θ, φ), it has the form

$$-\frac{\hbar^2}{2m}\left[\frac{\partial^2\psi}{\partial r^2}+\frac{2}{r}\frac{\partial\psi}{\partial r}-\frac{L^2}{r^2}\psi+\frac{e^2\psi}{r}\right]=E\psi \tag{8.7}$$

where

$$L^2=-\left[\frac{1}{\sin^2\theta}\frac{\partial^2}{\partial\varphi^2}+\frac{1}{\sin\theta}\frac{\partial}{\partial\theta}\left(\sin\theta\frac{\partial}{\partial\theta}\right)\right] \tag{8.8}$$

is a separation constant which equals the square of angular momentum.

In parabolic coordinates ξ,η,ϕ defined by

$$\begin{aligned} x &= \sqrt{\xi\eta}\,\cos\phi \\ y &= \sqrt{\xi\eta}\,\sin\phi \\ z &= \frac{1}{2}(\xi-\eta) \\ r &= \sqrt{x^2+y^2+z^2} = \frac{1}{2}(\xi+\eta) \end{aligned} \tag{8.9}$$

it has the form

$$\frac{4}{\xi+\eta}\left[\frac{\partial}{\partial\xi}\left(\xi\frac{\partial\psi}{\partial\xi}\right)+\frac{\partial}{\partial\eta}\left(\eta\frac{\partial\psi}{\partial\eta}\right)\right]+\frac{1}{\xi\eta}\frac{\partial^2\psi}{\partial\phi^2} + \frac{2m}{\hbar^2}\left(E+\frac{2e^2}{\xi+\eta}\right)\psi=0. \tag{8.10}$$

We shall first consider the solutions of the equation in spherical polar coordinates for both negative and positive and later in parabolic coordinates for negative energy eigenvalues only. Positive energy eigenvalue solutions for the latter will be taken up when we consider scattering of charged particles by a Coulomb field.

8.2 Solutions in Spherical Polar Coordinates

We shall use Coulomb units where the units of mass, length and time will be m, $\hbar^2/me^2$ and $\hbar^3/me^4$ respectively. This amounts to effectively putting $m=\hbar=e=1$. In these units the Schrödinger equation (8.7) reads as

$$\frac{\partial^2\psi}{\partial r^2}+\frac{2}{r}\frac{\partial\psi}{\partial r}-\frac{L^2}{r^2}\psi+2\left(E+\frac{1}{r}\right)\psi=0. \tag{8.11}$$

We have already shown in the previous sections that

$$\psi(r,\theta,\varphi)=R(r)Y_{lm}(\theta,\varphi), \tag{8.12}$$

which when substituted in (8.7) gives the radial equation

$$\frac{d^2R}{dr^2} + \frac{2}{r}\frac{dR}{dr} - \frac{l(l+1)}{r^2}R + 2\left(E + \frac{1}{r}\right)R = 0, \tag{8.13}$$

where l and m define the eigenvalues of L^2 and L_z;

$$L^2 Y_{lm} = l(l+1)Y_{lm} \qquad L_z Y_{lm} = mY_{lm}. \tag{8.14}$$

In terms of the new coordinates,

$$\rho = \frac{2r}{\alpha}, \quad \text{where} \quad \alpha = \frac{1}{\sqrt{-2E}}, \tag{8.15}$$

(8.13) has the form

$$R'' + \frac{2}{\rho}R' + \left[-\frac{1}{4} + \frac{\alpha}{\rho} - \frac{l(l+1)}{\rho^2}\right]R = 0. \tag{8.16}$$

Bound state

We shall seek solutions of this equation for $E < 0$ to obtain bound states whose radial wavefunctions at $\rho \to \infty$ must not diverge more rapidly than every finite power of ρ but remain finite at $\rho = 0$. For this we consider its limits for $\rho \to 0$ and $\rho \to \infty$. For $\rho \to 0$, the equation becomes

$$R'' + \frac{2}{\rho}R' = \frac{l(l+1)}{\rho^2}R. \tag{8.17}$$

Its solutions which are finite at $\rho = 0$ are

$$R(\rho) = \rho^l. \tag{8.18}$$

For $\rho \to \infty$, it reduces to

$$R'' = \frac{1}{4}R \tag{8.19}$$

whose solution that vanishes at infinity is

$$R = e^{-\rho/2}. \tag{8.20}$$

Taking into account the behaviours of R at $\rho \to 0$ and $\rho \to \infty$ as obtained above in (8.18) and (8.20), it is clear that $R(\rho)$ will have the form

$$R(\rho) = \rho^l e^{-\rho/2}\omega(\rho). \tag{8.21}$$

Substitution of this in (8.16) gives

$$\rho\omega''(\rho) + (2l + 2 - \rho)\omega'(\rho) + (\alpha - l - 1)\omega(\rho) = 0, \tag{8.22}$$

which is the confluent hypergeometric equation whose solution that satisfies the condition at $\rho = 0$ is

$$\omega(\rho) = F(-\alpha + l + 1, 2l + 2, \rho). \tag{8.23}$$

$F(a, b, \rho)$, the hypergeometric function above, is defined by the series

$$F(a, b, \rho) = 1 + \frac{a\rho}{b} + \frac{a(a+1)}{b(b+1)}\frac{\rho^2}{2!} + \cdots \tag{8.24}$$

which converges for all finite ρ with arbitrary a and non-zero, non-negative number b. However for $\rho \to \infty$, the series diverges as e^ρ and $R(\rho)$ would not satisfy the condition at infinity. For this to happen, the series must terminate and become a polynomial. This actually happens when a is either zero or a negative integer . For instance, when $a = -1$ the series has only two terms. When $a = -n$, where n is a positive integer, there are $(n + 1)$ terms in the series. It is related to the associated Lauguerre polynomial. In the solution (8.23) we must have

$$-\alpha + l + 1 = -n_r, \quad n_r = 0, 1, 2, \ldots \tag{8.25}$$

Since $\alpha = 1/\sqrt{-2E}$, (8.25) is equivalent to

$$E = -\frac{1}{2\alpha^2} = -\frac{1}{2(n_r + l + 1)^2} = -\frac{1}{2n^2}. \tag{8.26}$$

Since $\alpha = n = n_r + l + 1$

$$l = 0, 1, 2, \ldots (n - 1),$$

and n will have values $n = 1, 2, 3, \ldots$. The energy eigenvalues are independent of l which gives rise to what is termed as l-degeneracy or Coulomb degeneracy. Since there are $(2l + 1)$ values of m, the degree of degeneracy d works out to be

$$d = \sum_{l=0}^{n-1}(2l + 1) = n^2. \tag{8.27}$$

This l degeneracy is called accidental degeneracy because there is apparent lack of a corresponding symmetry associated with it. However, we shall see later in section 5 of this chapter that there is such a symmetry.

The normalised radial wave function when expressed in terms of r works out to

$$R_{nl}(r) = \frac{2(2r)^l e^{-r/n}}{n^{l+2}(2l+1)!}\sqrt{\frac{(n+l)!}{(n-l-1)!}}\, F\left(-n+l+1, 2l+2, \frac{2r}{n}\right)$$

$$= -\frac{2}{n^2}\frac{\sqrt{(n-l-1)!}}{(n+l)!} e^{-r/n}\left(\frac{2r}{n}\right)^l L_{n+l}^{2l+2}\left(\frac{2r}{n}\right), \tag{8.28}$$

where $L_{n+l}^{2l+2}(\frac{2r}{n})$ are the associated Laguerre polynomials, and the energy eigenfunctions become

$$\psi_{nlm}(r, \theta, \varphi) = R_{nl}(r) Y_{lm}(\theta, \varphi). \tag{8.29}$$

Continuum states

These states have positive energy whose spectrum is continuous, extending from $E = 0$ to $E = \infty$. Each of these states is infinitely degenerate. For each value of E there are infinite number of l values ranging from 0 to ∞.

Setting

$$E = \frac{k^2}{2} \tag{8.30}$$

in (8.15) and (8.26), we get

$$n = -i/k \quad \rho = 2ikr \tag{8.31}$$

so that the solutions of radial Schrödinger equation become

$$R_{kl} = \frac{C_{kl}(2kr)^l}{(2l+1)!} e^{-ikr} F\left(\frac{i}{k} + l + 1, 2l+2, 2ikr\right). \tag{8.32}$$

The normalisation C_{kl} is found to be

$$C_{kl} = 2ke^{\pi/2}|\Gamma(l+1-i/k)| \tag{8.33}$$

and

$$R_{kl} \underset{r\to\infty}{\longrightarrow} \frac{2}{r}\sin\left(kr + \frac{1}{k}\log 2kr - \frac{1}{2}l\pi + \delta_l\right),$$

$$\delta_l = \arg\Gamma\left(l + 1 - \frac{i}{k}\right). \tag{8.34}$$

Repulsive Coulomb potential

Instead of $V = -\alpha/r$ if we have $V = \alpha/r$, there will be no bound states. The radial Schrödinger equation for this case can be obtained from that of the attractive potential by changing the sign of r. The eigenfunctions are thus

$$R_{kl} = \frac{C_{kl}}{(2l+1)!}(2kr)^l e^{ikr} F(i/k + l + 1, 2l + 1, -2ikr) \quad (8.35)$$

$$\text{with} \quad C_{kl} = 2ke^{-\pi/2k}\left|\Gamma\left(l + 1 + \frac{i}{k}\right)\right| \quad (8.36)$$

$$\text{and} \quad R_{kl} \xrightarrow[r\to\infty]{} \frac{2}{r}\sin\left(kr - \frac{1}{k}\log 2kr - \frac{1}{2}l\pi + \delta\right),$$

$$\text{with} \quad \delta_l = \arg\Gamma\left(l + 1 + \frac{i}{k}\right). \quad (8.37)$$

8.3 Solutions in Parabolic Coordinates

An inspection of the Schrödinger equation in parabolic coordinates given in (8.10) reveals that the wavefunction $\psi(\xi, \eta, \phi)$ is separable in ξ, η and ϕ. We shall therefore try with

$$\psi(\xi, \eta, \phi) = f_1(\xi) f_2(\eta) e^{im\varphi}. \quad (8.38)$$

Substitution of this in (8.10) gives

$$\frac{d}{d\xi}\left(\xi\frac{df_1}{d\xi}\right) + \left(\frac{1}{2}E\xi - \frac{m^2}{4\xi} + \beta_1\right) f_1 = 0, \quad (8.39a)$$

$$\frac{d}{d\eta}\left(\eta\frac{df_2}{d\eta}\right) + \left(\frac{1}{2}E\eta - \frac{m^2}{4\eta} + \beta_2\right) f_1 = 0, \quad (8.39b)$$

with

$$\beta_1 + \beta_2 = 1, \quad (8.39c)$$

β_1, β_2 being the separation constants.

In terms of the variables

$$\rho_1 = \xi\sqrt{-2E} = \frac{\xi}{\alpha}, \quad \rho_2 = \eta\sqrt{-2E} = \frac{\eta}{\alpha}, \quad \alpha = \frac{1}{\sqrt{-2E}}, \quad (8.40)$$

(8.39a) and (8.39b) have the forms

$$\frac{d^2 f_1}{d\rho_1^2} + \frac{1}{\rho_1}\frac{df_1}{d\rho_1} + \left[-\frac{1}{4} + \frac{1}{\rho_1}\left(\frac{|m|+1}{2} + n_1\right) - \frac{m^2}{4\rho_1^2}\right] f_1 = 0, \quad (8.41a)$$

$$\frac{d^2 f_2}{d\rho_2^2} + \frac{1}{\rho_2}\frac{df_2}{d\rho_2} + \left[-\frac{1}{4} + \frac{1}{\rho_2}\left(\frac{|m|+1}{2} + n_2\right) - \frac{m^2}{4\rho_2^2}\right] f_2 = 0, \quad (8.41b)$$

where

$$n_1 = -\frac{1}{2}(|m|+1) + \alpha\left(\frac{1-\beta}{2}\right),$$
$$n_2 = -\frac{1}{2}(|m|+1) + \alpha\left(\frac{1+\beta}{2}\right), \quad (8.41c)$$

with $\beta_1 = \frac{1}{2}(1-\beta)$, $\beta_2 = \frac{1}{2}(1+\beta)$.
Following the procedure of the previous section we find that

$$f_1(\rho_1) \underset{\rho_1 \to 0}{\longrightarrow} (\rho_1)^{\frac{1}{2}|m|}, \quad f_1(\rho_1) \underset{\rho_1 \to \infty}{\longrightarrow} e^{-\frac{1}{2}\rho_1}. \quad (8.43)$$

We therefore put

$$f_1(\rho_1) = \rho_1^{\frac{1}{2}|m|} e^{-\frac{1}{2}\rho_1} \omega_1(\rho_1) \quad (8.44)$$

the substitution of which in (8.41a) gives

$$\rho_1 \omega_1''(\rho_1) + (|m| + 1 - \rho_1)\omega_1'(\rho_1) + n_1 \omega_1(\rho_1) = 0. \quad (8.45)$$

This is a confluent hypergeometric equation with solution

$$\omega_1(\rho_1) = F(-n_1, |m|+1, \rho_1). \quad (8.45a)$$

For ω_1 to be polynomial bound at infinity, n_1 has to be a non-negative integer including zero. A similar procedure gives

$$\omega_2(\rho_2) = F(-n_2, |m|+1, \rho_2), \quad (8.45b)$$

where n_2 is a non-negative integer including zero.
Hence

$$n_1 + n_2 = 0, 1, 2, \ldots . \quad (8.46)$$

Now from (8.41c) we have

$$n_1 + n_2 + |m| + 1 = \alpha = \frac{1}{\sqrt{-2E}}, \quad (8.47)$$

from which we obtain energy eigenvalues

$$E = -\frac{1}{2(n_1 + n_2 + |m| + 1)^2}. \tag{8.48}$$

Comparing with (8.26), we find

$$n = n_1 + n_2 + |m| + 1. \tag{8.49}$$

It may be noted that E is independent of the separation constant β, similar to the spherical polar coordinate case where the energy is independent of separation constant $L^2 = l(l+1)$. Its value obtained from (8.41c) is

$$\beta = \left(\frac{n_1 - n_2}{n}\right). \tag{8.50}$$

This constant can be shown to be the third component of the Laplace–Runge–Lenz vector $\vec{A}$ appearing later in this chapter in section 4. The energy eigenvalue can be expressed in terms of β:

$$E = -\frac{\beta^2}{2(n_1 - n_2)^2}. \tag{8.51}$$

It is to be noted that $(n_1 - n_2)$ is known as 'electric quantum number'.

Since for a given n, $|m|$ can take value from 0 to $n-1$ and for fixed n and $|m|$, n_1 has values from 0 to $n = |m| - 1$, the degree of degeneracy for a state with quantum number n can be calculated as

$$d = 2\sum_{m}^{n-1}[(n-m)+n] = n^2. \tag{8.52}$$

Here $m = \pm|m|$. The normalised wavefunctions as collected from (8.38) and (8.45) work out to be

$$\psi_{n,n_1n_2m}(\xi,\eta,\phi) = \frac{\sqrt{2}}{n^2} f_{n_1m}\left(\frac{\xi}{n}\right) f_{n_2m}\left(\frac{\eta}{n}\right) \frac{e^{im\varphi}}{\sqrt{2\pi}}, \tag{8.53}$$

where

$$f_{Nm}(\rho) = \frac{e^{-\rho/2}\rho^{\frac{1}{2}|m|}}{m!}\left[\frac{N+|m|!}{N!}\right]^{1/2} F(-N, |m|+1, \rho). \tag{8.54}$$

Since $n = n_1 + n_2 + m + 1$, it is enough to characterise the wavefunction by quantum numbers n_2, n_1 and m only. These wavefunctions $\psi_{n,n_1,n_2}(\xi,\eta,\phi)$ in parabolic coordinates can be expressed as linear combinations of the corresponding wavefunction in spherical polar coordinates. For the states $n = 2$, these relations are

$$\psi_{200}(r,\theta,\phi) = \frac{1}{\sqrt{2}}\left[\psi_{201}(\xi,\eta,\phi) + \psi_{210}(\xi,\eta,\phi)\right],$$

$$\psi_{210}(r,\theta,\phi) = \frac{1}{\sqrt{2}}\left[\psi_{201}(\xi,\eta,\phi) - \psi_{210}(\xi,\eta,\phi)\right]. \tag{8.55}$$

8.4 Origin of Coulomb Degeneracy

Whenever a system has a symmetry, associated with it is a degeneracy of the states of the system. For instance, if the Hamiltonian of the system is invariant under rotation, components of the angular momentum operator $\vec{L}$ are conserved and the energy eigenvalues are independent of those of L_3, i.e. m, the magnetic quantum number, which is a case of m-degeneracy. We have seen that in the case of hydrogen atom there is an additional degeneracy: the energy eigenvalues are independent of l which we have designated as l-degeneracy. It is also known as Coulomb degeneracy because it is very specific to the Coulomb potential. It transpires that there is an underlying symmetry behind this degeneracy. It has been known for quite some time that in addition to conservation of the angular momentum $\vec{L}$, the Coulomb potential has an additional conserved vector $\vec{A}$ known as Laplace–Runge–Lenz vector

$$\vec{A} = \frac{\vec{r}}{r} - \vec{p}\times\vec{L}, \tag{8.56}$$

which is orthogonal to $\vec{L}$ and whose magnitude in classical mechanics is the eccentricity of the orbit of a charged particle moving in a Coulomb potential. In quantum mechanics, we shall use a symmetrised and properly normalised operator $\vec{M}$ in place of $\vec{A}$:

$$\vec{M} = \left[\frac{\vec{r}}{r} - \frac{1}{2}\left(\vec{p}\times\vec{L} - \vec{L}\times\vec{p}\right)\right]\frac{1}{\sqrt{-2E}}. \tag{8.57}$$

It commutes with the Hamiltonian and its components obey commutation relations

$$[M_i, M_j] = i\epsilon_{ijk}L_k,$$
$$[L_i, M_j] = i\epsilon_{ijk}M_k, \tag{8.58}$$

which together with

$$[L_i, L_j] = i\epsilon_{ijk}L_k \tag{8.59}$$

constitute a $O(4)$ algebra and defines the enlarged symmetry of the hydrogen atom known as the $O(4)$ symmetry. It is a direct product of two independent algebras defined by

$$
\begin{aligned}
&[J_i^{(+)}, J_j^{(+)}] = i\epsilon_{ijk} J_k^{(+)} \\
&[J_i^{(-)}, J_j^{(-)}] = i\epsilon_{ijk} J_k^{(-)} \\
&[J_i^{(+)}, J_j^{(-)}] = 0,
\end{aligned}
\tag{8.60}
$$

where

$$J_i^{(\pm)} = \frac{1}{\sqrt{2}}(L_i \pm M_i). \tag{8.61}$$

Since the components of $\vec{J}^{(+)}$ and $\vec{J}^{(-)}$ obey algebras (8.60), we have

$$(\vec{J}^{(+)})^2 = (\vec{J}^{(-)})^2 = j(j+1). \tag{8.62}$$

The energy eigenvalues of the hydrogen atom can be calculated without solving the Schrödinger equation from the above algebraic relations and the identity

$$\vec{L}^2 + \vec{M}^2 = -1 - \frac{1}{2E}, \tag{8.63}$$

from which we get

$$(\vec{J}^{(-)})^2 = (\vec{J}^{(+)})^2 = j(j+1)) = \vec{L}^2 + \vec{M}^2 = -1 - \frac{1}{2E} \tag{8.64}$$

leading to

$$E = -\frac{1}{2(2j+1)^2} = -\frac{1}{2n^2}, \tag{8.65}$$

where

$$n = 2j + 1\ . \tag{8.66}$$

Since $n = 1, 2, 3, \ldots, j = 0, \frac{1}{2}, 1, \frac{3}{2}, \ldots,$ which shows that the algebras occuring in (8.60) are the SU(2) algebras.

Problems

1. Show that the separation constant β appearing in the solution of Schrödinger equation in parabolic coordinates is the third component of the Laplace–Runge–Lenz vector.
   Hint: In parabolic coordinates,

$$
\begin{aligned}
A_3 = &\frac{2\xi\eta}{\xi+\eta}\left(\frac{\partial^2}{\partial\eta^2} + \frac{1}{\eta}\frac{\partial}{\partial\eta} - \frac{\partial^2}{\partial\xi^2} - \frac{1}{\xi}\frac{\partial}{\partial\xi} - \frac{m^2}{2n}\frac{\xi-\eta}{\xi\eta}\right) \\
&+ \frac{\xi-\eta}{\xi+\eta}.
\end{aligned}
$$

   Solve for $A_3\psi_{n_1,n_2,m}(\xi,\eta,\phi) = \lambda\psi_{n_1 n_2}(\xi,\eta,\phi)$ and obtain $\lambda = \frac{n_1-n_2}{n}$.

2. Find the probability distribution of proton and electron separately as a function of distance from their centre of mass for the ground states of the hydrogen atom.
   Hint: Take the centre of mass as the origin and obtain proton and electron coordinates in terms of relative coordinates and obtain the probability density $D = 4\pi r^2 |R(r)|^2$.

3. A hydrogen atom is in the $2p$ state. What is the probability that the electron is less distant from the proton than the Bohr radius?
   Hint: There are three states ($m = 1, 0 - 1$). Obtain $P(a_0)$ by restricting the integration over the relative coordinate to a_0.

4. The proton in the hydrogen atom ground state is displaced through a distance b. What is the probability of the atom being in the ground state?
   Hint: The required result is the volume integral over the product of the ground state wave functions $\psi^*(\vec{r}_1)$ and $\psi(\vec{r}_2)$ where $\vec{r}_1$ is relative coordinate with the centre of mass at the origin of the undisplaced atom and $\vec{r}_2$ that of the displaced atom. Use spheroidal coordinates: $\vec{u} = \vec{r}_1 + \vec{r}_2$, $\vec{v} = \vec{r}_1 - \vec{r}_2$ and φ defined by $x = \frac{uv}{2b}$ $y = r\cos\phi$, $z = r\sin\phi$, where $\vec{r}$ is radius vector of the electron from the middle of the line joining the original displaced origins.

   Answer: Probability $= \left(\frac{b^2+3a_0b+3a_0^2}{3a_0^2}\right)^2 e^{-2b/a_0}$

Bibliography

[1] W. Lenz, *Zeit. f. Physik*, **24**, 197 (1924).
[2] W. Pauli, *Zeit. f. Physik*, **36**, 336 (1926).
[3] L.D. Landau and E.M. Lifshitz, *Quantum Mechanics*, 3rd edition, Pergamon Press, Oxford (1977).
[4] L.I. Schiff, *Quantum Mechanics*, 3rd edition, McGraw Hill Book, Inc. (1977)
[5] V. Fock, *Z. Phys.*, **98**, 145 (1935).
[6] H.V. McIntosh, *Am. J. Phys.*, **27**, 620 (1939).
[7] S.P. Alliluev, *Soviet Phys. JETP*, **6**, 156 (1958).
[8] A.A. Sokolov, I.M. Ternov and V.Ch. Zhukovskii, *Quantum Mechanics*, Mir Publishers, Moscow (1984).

9 Action in Quantum Mechanics

9.1 Introduction

Planck and Einstein resolved the problems posed by black-body radiation and photoelectric effect by quantizing energy while Bohr and Sommerfeld explained discrete spectra of atoms by quantizing 'action'. Although energy and angular momentum play central role in the quantum mechanics of Heisenberg, Schrödinger and Dirac, the quantum numbers n and l that appear in the treatment of hydrogen and other atoms are the quanta of Bohr–Sommerfeld 'action'. Not much attention has been paid to the construction of quantum mechanical operators that represent these action functions and the solution of the eigenvalue equations to obtain their eigenfunctions. Some of these operators exist in literature without being identified as such. In this chapter we give systematic presentation of the quantum mechanics of 'action'.

9.2 Azimuthal Action Operator

In classical mechanics, the square of the angular momentum vector is related to the action function S_a by the Hamilton–Jacobi differential equation

$$\vec{L}^2(\theta, \varphi) = \left(\frac{\partial S_a}{\partial \theta}\right)^2 + \frac{1}{\sin^2 \theta}\left(\frac{\partial S_a}{\partial \varphi}\right)^2, \tag{9.1}$$

whose solution, when quantized, yields

$$S_a = l, \quad l = 0, 1, 2, \ldots . \tag{9.2}$$

We label the quantum number l as the 'azimuthal quantum number', which differs from Sommerfeld quantum number by one unit. In quantum mechanics, this quantum number appears in the eigenvalue equation for the operator $\vec{L}^2$:

$$\begin{aligned}\vec{L}^2 Y_{l,m}(\theta, \varphi) &= -\left[\frac{\partial}{\sin\theta\partial\theta}\left(\sin\theta\frac{\partial}{\partial\theta}\right) + \frac{1}{\sin^2\theta}\frac{\partial^2}{\partial\varphi^2}\right] Y_{l,m}(\theta, \varphi)\\ &= l(l+1)Y_{l,m}(\theta, \varphi).\end{aligned} \tag{9.3}$$

While seeking to obtain the form of the operator for S_a which has eigenvalue l, we note that it should be related to the operator $\vec{L}^2$ by

$$\vec{L}^2 = S_a(S_a + 1). \tag{9.4}$$

From (9.3) and (9.4) it follows that such an operator would have two eigenvalues, namely l and $-(l+1)$:

$$\begin{aligned} S_a\Omega_{l,m}(\theta,\varphi) &= l\Omega_{l,m}(\theta,\varphi), \\ S_a\Omega_{-(l+1),m}(\theta,\varphi) &= -(l+1)\Omega_{-(l+1),m}(\theta,\varphi). \end{aligned} \tag{9.5}$$

Equation (9.4) has the solution

$$S_a = -\frac{1}{2} + \frac{1}{2}\sqrt{1+4L^2} = \vec{\sigma}\cdot\vec{L} \tag{9.6}$$

since $\sqrt{1+4l^2} = 1 + 2\vec{\sigma}\cdot\vec{L}$. Here σ_1, σ_2 and σ_3 are the Pauli spin matrices. The operator $S_a = \vec{\sigma}\cdot\vec{L}$ is a matrix of the form

$$S_a = \begin{pmatrix} L_3 & L_{(-)} \\ L_{(+)} & -L_3 \end{pmatrix}. \tag{9.7}$$

It is easy to see that this operator has two eigenvalues, l and $-(l+1)$, as required. The eigenfunctions corresponding to these two eigenvalues work out to be (see Appendix B for derivation)

$$\Omega_{l,m}(\theta\varphi) = \begin{pmatrix} \sqrt{\frac{l+m+1}{2l+1}} & Y_{l,m}(\theta,\varphi) \\ \sqrt{\frac{l-m}{2l+1}} & Y_{l,m+1}(\theta,\varphi) \end{pmatrix}, \tag{9.8a}$$

$$\Omega_{-(l+1),m}(\theta\varphi) - \begin{pmatrix} -\sqrt{\frac{l-m+1}{2l+1}} & Y_{l,m-1}(\theta,\varphi) \\ \sqrt{\frac{l+m}{2l+1}} & Y_{l,m}(\theta,\varphi) \end{pmatrix}, \tag{9.8b}$$

which are known as 'spinor spherical harmonics'. We have seen in section 5.7 that these are also eigenfunctions of a free spin-1/2 particle having $j = l + \frac{1}{2}$ and $j = l - \frac{1}{2}$ respectively. This interesting result arises from the linearization, as obtained in (9.6), of the nonlinear relation

$$S_a^2 + S_a = \vec{L}^2, \tag{9.9}$$

which is analogous to Dirac's linearisation

$$E = \vec{\alpha}\cdot\vec{p} + \beta m \tag{9.10a}$$

of Einstein's nonlinear formula

$$E^2 = p^2 + m^2, \tag{9.10b}$$

where the linear relation (9.10a) leads to the spin of relativistic particles.

A further interesting result comes from

$$[S_a, \vec{L}^2] = 0. \tag{9.11}$$

This commutability tells us that S_a and $\vec{L}^2$ have common eigenfunctions. It is clear that $Y_{lm}(\theta\varphi)$ is not this one. One can check that $\Omega_{lm}(\theta\varphi)$ given in (9.8a) is the common eigenfunction. In this connection it is to be noted that since, for a spherical top,

$$H = \frac{\vec{L}^2}{2I}, \tag{9.12}$$

I being the moment of inertia about the axis of rotation, action S_a commutes with the Hamiltonian:

$$[S_a, H] = 0. \tag{9.13}$$

As a result, $\Omega_{l,m}(\theta\varphi)$ is the common eigenfunction of S_a and H. This is analogous to the situation in relativistic quantum mechanics where the eigenfunctions of $H = \vec{\alpha} \cdot \vec{p} + \beta m$ are also the eigenfunctions of H^2, but not vice versa, which leads to the conclusion that spin is associated with the relativistic in quantum mechanics. It is therefore tempting to conclude that a quantum mechanical spherical top can have spin-1/2 in addition to spin-zero and even go one step further to conclude the same for all spherically symmetric quantum mechanical systems whose Hamiltonian commutes with $\vec{L}$ and therefore with S_a.

A systematic diagonalisation of a matrix of the form (9.7), where the elements are not hermitian, was performed by Dirac in the context of relativistic wave equations. We follow his method for the present case where all the matrix elements of S_a are hermitian, as a result of which it can be diagonalised by a unitary matrix U:

$$U^+ S_a U = D, UU^+ = U^+ U = 1, \tag{9.14}$$

where D is the diagonal matrix. The unitary matrix is constructed from the eigenfunctions of S_a:

$$U = \frac{1}{\sqrt{2l+1}} \begin{pmatrix} \omega_1 & v_1 \\ \omega_2 & v_2 \end{pmatrix}, \tag{9.15}$$

$$S_a v = lv \quad S_a \omega = -(l+1)\omega \quad , v = \begin{pmatrix} v_1 \\ v_2 \end{pmatrix} \quad , \quad \omega = \begin{pmatrix} \omega_1 \\ \omega_2 \end{pmatrix}. \tag{9.16}$$

Here v_1 and v_2 are the rectangular matrices having $(2l+1)$ rows and $2l$ columns while ω_1 and ω_2 have $(2l+1)$ rows and $(2l+2)$ columns. The unitarity conditions give

$$v^+ v = 2l + 1\,, \quad vv^+ = 2l. \tag{9.17}$$

Multiplication of the first equation by v_2^+ on the right and the second by the same on the left, followed by subtraction gives

$$v_1^+ v_1 v_2^+ - v_2^+ v_1 v_1^+ = v_2^+. \tag{9.18a}$$

Following the same method gives the following three more relations:

$$v_1^+ v_2 v_2^+ - v_2^+ v_2 v_1^+ = v_1^+ \tag{9.18b}$$

$$v_1 v_1^+ v_2 - v_2 v_1^+ v_1 = -v_2 \tag{9.18c}$$

$$v_1 v_2^+ v_2 - v_2 v_2^+ v_1 = v_1 \tag{9.18d}$$

These trilinear commutation relations happen to be those of parabose algebra of order 2. Similar relations can be obtained for ω_1 and ω_2. It is worth pondering about the significance of this result.

9.3 Action Operator for Spin Angular Momentum

We shall represent the spin angular momentum J_i as

$$J_i = \frac{1}{2} a^+ \sigma_i a, \quad a = \begin{pmatrix} a_1 \\ a_2 \end{pmatrix}, \tag{9.19}$$

where a_α^+ and a_α $(\alpha = 1, 2)$ are the creation–annihilation operators obeying the commutation rules

$$[a_\alpha, a_\beta^+] = \delta_{\alpha\beta}\ , \quad [a_\alpha, a_\beta] = 0\ , \ [a_\alpha^+, a_\beta^+] = 0. \tag{9.20}$$

From (9.19) we have

$$\vec{J}^2 = \frac{1}{4} (\upsilon_i)_{\alpha\beta} (\upsilon_i)_{\lambda n} a_\alpha^+ a_\beta a_\lambda^+ a_n, \tag{9.21}$$

which can be simplified by using the identity

$$((\alpha_i)_{\alpha\beta} (\sigma_i)_{\lambda n}) = 2\delta_{\alpha n}\delta_{\beta\lambda} - \delta_{\alpha\beta}\delta_{\lambda n} \tag{9.22}$$

and the commutation relation (9.20). We get

$$\vec{J}^2 = \frac{1}{4} (a_\alpha^+ a_\alpha)^2 + \frac{1}{2} (a_\alpha a_\alpha) = S(S+1) \tag{9.23}$$

with

$$S = \frac{1}{2} a_\alpha^+ a_\alpha . \tag{9.24}$$

Since

$$\vec{J}^2 |j, m\rangle = j(j+1)|j, m\rangle, \tag{9.25}$$

it follows from (9.23) and (9.24) that

$$S|j, m\rangle = j|j, m\rangle \tag{9.26}$$

with

$$j = \frac{1}{2} a_\alpha^+ a_\alpha = \frac{1}{2} n \ , \ n = 0, 1, 2, \ldots . \tag{9.27}$$

The eigenfunction corresponding to the eigenvalue j can be represented as

$$|j, m\rangle = \frac{(a_1^+)^{j+m} (a_2^+)^{j-m}}{\sqrt{(j+m)!(j-m)!}} |0, 0\rangle, \tag{9.28}$$

where the state $|0, 0\rangle$ is defined by

$$a_\alpha |0, 0\rangle = 0. \tag{9.29}$$

We thus have the operator form of S given by (9.24) and its eigenfunction is (9.28). It may noted that this S is a scalar operator whereas the azimuthal action operator of the previous section is a scalar matrix, which for the present case has the form

$$\begin{pmatrix} \frac{1}{2}(a_1^+ a_1 - a_2^+ a_2) & a_1^+ a_2 \\ a_2^+ a_1 & \frac{1}{2}(a_2^+ a_2 - a_1^+ a_1) \end{pmatrix} .$$

9.4 Principal Action Operator for H-atom

In classical mechanics, the action function S for the hydrogen atom satisfies the Hamilton–Jacobi equation

$$\frac{1}{2} (\vec{\nabla} S)^2 = E + \frac{1}{r} \ , \ (e = m = 1), \tag{9.30}$$

which, in spherical polar coordinates becomes

$$\frac{1}{2}\left[\left(\frac{\partial S}{\partial r}\right)^2 + \frac{1}{r^2}\left(\frac{\partial S}{\partial \theta}\right)^2 + \frac{1}{r^2 \sin^2\theta}\left(\frac{\partial S}{\partial \varphi}\right)^2\right] = E + \frac{1}{r} . \tag{9.31}$$

This equation is separable:

$$S = S_r(r) + S_a(\theta\varphi) \tag{9.32}$$

$$\frac{1}{r^2}\left[\left(\frac{\partial S_r}{\partial r}\right)^2\right] = 2\left(E + \frac{1}{r}\right) - \frac{\vec{L}^2}{r^2} \tag{9.33}$$

$$\left[\left(\frac{\partial S_a}{\partial \theta}\right)^2 + \frac{1}{\sin^2\theta}\left(\frac{\partial S_a}{\partial \varphi}\right)^2\right] = \vec{L}^2 \tag{9.34}$$

Integration of these two equations followed by imposition of Bohr–Sommerfeld quantization leads to

$$S_r = n_r \qquad S_a = l\ , \tag{9.35}$$

with

$$n_r + l = n \tag{9.36}$$

so that

$$S = n. \tag{9.37}$$

In quantum mechanics, we would have,

$$S\phi_{n,l,m}(\vec{r}) = n\phi_{n,l,m}(\vec{r}) \tag{9.38}$$

$$H\psi_{n,l,m}(\vec{r}) = \left(\frac{p^2}{2} - \frac{1}{r}\right)\psi_{n,l,m}(\vec{r}) = E_n\psi_{n,l,m}(\vec{r}). \tag{9.39}$$

In order to find the operator form of S, we make the ansatz that

$$\psi_{n,l,m}(\vec{r}) = U\phi_{n,l,m}(\vec{r})\ ,\quad UU^+ = U^+U = 1. \tag{9.40}$$

Substituting (9.40) in (9.38), we get

$$USU^{-1}\psi_{n,l,m}(\vec{r}) = n\psi_{n,l,m}(\vec{r}). \tag{9.41}$$

Expressing the unitary operator U in terms of a hermitian operator T,*

$$U = e^{-i\delta T},\quad T^{\dagger} = T, \tag{9.42}$$

and then substituting in (9.41), we get

$$\left\{S - i\delta\,[T, S] + \frac{(-i\delta)^2}{2!}[T, [T, S]] + \cdots\right\}\psi_{n,l,m}(\vec{r}) = n\psi_{n,l,m}(\vec{r}). \tag{9.43}$$

---

*See footnote on page 99.

Now, if

$$[T, S] = iW \quad \text{and} \quad [T, W] = iS, \tag{9.44}$$

then (9.43) can be written as

$$\left\{\frac{1}{2}(S + W)e^{\delta} + \frac{1}{2}(S - W)e^{-\delta}\right\} \psi_{n,l,m}(\vec{r}) = n\psi_{n,l,m}(\vec{r}). \tag{9.45}$$

On the other hand, (9.39) can be put in the form

$$\left(\frac{rp^2}{2}\frac{1}{\sqrt{-2E_n}} - \frac{r}{2}\sqrt{-2E_n}\right) \psi_{n,l,m}(\vec{r}) = \frac{1}{\sqrt{-2E_n}}\psi_{n,l,m}(\vec{r}). \tag{9.46}$$

A comparison of (9.45) and (9.46) gives

$$(S + W) = rp^2\,, \quad (S - W) = r \tag{9.47}$$

$$\text{if} \quad e^{\delta} = \frac{1}{\sqrt{-2E_n}} = n, \tag{9.48}$$

from which we get

$$S = \frac{1}{2}(rp^2 + r)\,, \quad W = \frac{1}{2}(rp^2 - r)\,, \quad E_n = -\frac{1}{2n^2}. \tag{9.49}$$

We thus get the quantization of energy from quantization of action as in Bohr's theory. In Schrödinger quantum mechanics, this is obtained from convergence of radial wave function at infinity, which requires termination of the relevant hypergeometric series. Putting

$$W = S_1, \quad T = S_2, \quad S = S_3, \tag{9.50}$$

commutators (9.44) become

$$[S_1 S_2] = -iS_3\,, \ [S_2, S_3] = iS_1. \tag{9.51}$$

Further, from (9.49), the commutator $[S_3, S_1]$ can be calculated and is found to be

$$[S_3 S_1] = \left(1 + r\frac{\partial}{\partial r}\right) = iS_2. \tag{9.52}$$

Equation (9.51) and (9.52) together represent $O(2, 1)$ algebra

$$[S_1 S_2] = -iS_3\,, \ [S_2 S_3] = iS_1\,, \ [S_3 S_1] = iS_2 \tag{9.53}$$

with

$$S_1 = \frac{1}{2}(rp^2 - r)\,, \quad S_2 = -i\left(1 + r\frac{\partial}{\partial r}\right)\,, \quad S_3 = \frac{1}{2}(rp^2 + r). \tag{9.54}$$

9.5 Solution of Eigenvalue Equation for Action for H-atom

The eigenvalue equation for the action operator $S = S_3$

$$S\phi_{n,l,m}(\vec{r}) = \frac{1}{2}(r - r\nabla^2)\phi_{n,l,m}(\vec{r}) = n\phi_{n,l,m}(\vec{r}) \tag{9.55}$$

is separable in spherical polar coordinates:

$$\phi_{n,l,m}(\vec{r}) = \mathcal{R}_{n,l}(r)\Omega_{l,m}(\theta,\varphi), \tag{9.56}$$

where $\Omega_{l,m}(\theta\varphi)$, the spherical harmonic spinor, is the eigenfunction of the azimuthal action operator. Substitution of this in (9.55) gives

$$\mathcal{R}''_{n,l}(r) + \frac{2}{r}\mathcal{R}'_{n,l}(r) + \left[\frac{2n}{r} - 1 - \frac{l(l+1)}{r^2}\right]\mathcal{R}_{n,l}(r) = 0. \tag{9.57}$$

This equation is solved in the manner in which the radial equation for energy eigenfunction is done by setting

$$\mathcal{R}_{n,l}(r) = N_{n,l}r^l e^{-r} f(r) \tag{9.58}$$

and substituting it in (9.57), which gives

$$rf''(r) + 2(l+1-r)f'(r) + 2(n-l-1)f(r) = 0, \tag{9.59}$$

which, in terms of $\rho = 2r$ reads

$$\rho f''(\rho) + (2l+2-\rho)f'(\rho) + (n-l-1)f(\rho) = 0. \tag{9.60}$$

This is a confluent hypergeometric equation whose solution is

$$f(\rho) = F(-n+l+1, 2l+2, \rho). \tag{9.61}$$

Substitution of this in (9.56) and (9.58) gives

$$\phi_{n,l,m}(\vec{r}) = N_{n,l}\Omega_{l,m}(\theta,\varphi(2r)^l e^{-r} F(-n+l+1, 2l+2, 2r), \tag{9.62}$$

The normalisation factor $N_{l,m}$ is to be determined from the orthonormality relation*

$$\int d^3r\phi^*_{n,l,m}(\vec{r})\frac{1}{r}\phi_{n',l',m'}(\vec{r}) = \delta_{m,m'}\delta_{l,l'}\delta_{m,m'}; \tag{9.63}$$

---

*It may be noted that the operator S does not satisfy the hermiticity condition (2.2) of chapter 2 but the modified condition

$$\int d\vec{r}\psi^*\frac{1}{r}(S\psi) = \int d\vec{r}(S\psi)^*\frac{1}{r}\psi.$$

it can be found to be

$$N_{n,l} = 2\sqrt{\frac{(n-l-1)!}{(n+l)!}}.$$

The function $\phi_{n,l,m}(\vec{r})$ is known as Coulomb Sturmian function, which resembles the energy eigenfunction.

$$\psi_{n,l,m}(\vec{r}) = \frac{N_{nl}}{n^2} Y_{lm}(\theta, \varphi) \left(\frac{2r}{n}\right)^l e^{-r/n} F\left(-n+l+1, 2l+2, \frac{2r}{n}\right). \tag{9.64}$$

It will be noticed that

$$\psi_{n,l,m}(\vec{r}) = \frac{1}{n^2}\phi_{n,l,m}\left(\frac{\vec{r}}{n}\right), \tag{9.65}$$

which is a reflection of the fact that

$$e^{-i\theta S_2}\vec{r}e^{i\theta S_2} = \frac{\vec{r}}{n}. \tag{9.66}$$

Problems

1. The action operator S for a particle moving in a potential

$$V = -\frac{e^2}{r} + \frac{g^2}{r^2}$$

is given to be

$$S = \frac{r}{\sqrt{2}}\left[\frac{1}{2} - \frac{1}{r^2}\frac{\partial}{\partial r}\left(r^2\frac{\partial}{\partial r}\right) + \frac{\Lambda^2}{r^2}\right],$$

where $\Lambda^2 = \lambda(\lambda+1) = l(l+1) + 2mg^2$. Show that its eigenvalues are

$$S = p + \lambda \quad , \quad p = 0, 1, 2, \ldots$$

and that of the Hamiltonian are

$$E = -\frac{me^4}{2(p+\lambda)^2}.$$

2. Show that the action operator in problem 1 does not commute with the Hamiltonian and this is why the energy eigenvalues depends on l.

3. Obtain the eigenfunctions of the principal action operator in parabolic coordinates.

Bibliography

[1] M. Bednar, *Ann. Phys.*, (N.Y.) **75**, 305 (1973).

[2] E.R. Vrcay, *Phys. Rev.*, A**31**, 2054 (1985).

[3] C.C. Gerry and J. Kiefer, *Phys. Rev.*, A**37**, 665 (1988).

[4] P.A.M. Dirac, *Proc. Roy. Soc.*, A**155**, 447 (1936).

[5] M. Rotenberg, *Ann. Phys.*, (N.Y.) **19**, 262 (1962).

[6] J. Schwinger, "On angular momentum" in *Quantum Theory of Angular Momentum* by L.C. Biedenharn and H. Van Dam, Academic Press, New York (1965) p. 229.

[7] B. Kursunoglu, *Modern Quantum Theory*, W.H. Freeman and Company, London (1962).

[8] H.C. Corben and P. Stehle, *Classical Mechanics*, John Wiley and Sons, Inc., New York (1950).

[9] V.A. Fock, *Fundamentals of Quantum Mechanics*, Mir Publishers, Moscow (1978).

10 Perturbation Theory

10.1 Introduction

There are a limited number of quantum mechanical systems of physical interest for which Schrödinger equation can be solved exactly. We have dealt with such systems in chapter 7. On the other hand, there are a number of such systems for which this can be done if a part of the Hamiltonian is omitted. This part is called the perturbation and if it happens to be such that its contribution to the energy eigenvalue is small, an approximation scheme can be developed to calculate it. This has come to be known as perturbation theory which is outlined below.

$$H = H_0 + H_I, \tag{10.1}$$

where the equation for the unperturbed Hamiltonian

$$H_0 \psi_n^{(0)} = E_n^{(0)} \psi_n^{(0)} \tag{10.2}$$

can be solved exactly and the perturbation Hamiltonian,

$$H_I = \lambda V, \tag{10.3}$$

is characterised by a small parameter λ. The 'potential' V can either be time-independent or time-dependent. The unperturbed Hamiltonian may have eigenstates which have degeneracies. We have to deal with these cases separately. We shall first take time-independent perturbation for non-degenerate states and then for degenerate states. Finally, we shall develop methods for time-dependent perturbations.

10.2 Time-Independent Perturbation for Non-Degenerate States

Rayleigh–Schrödinger method

The problem here is to find the energy eigenvalues E_s and the eigenfunctions ψ_s defined by

$$H\psi_s = (H_0 + \lambda V)\psi_s = E_s \psi_s. \tag{10.4}$$

Since the eigenfunction $\psi_n^{(0)}$ of H_0 form an orthonormal complete set, we expand ψ_s in terms of this set:

$$\psi_s(x) = \sum_n c_{ns}\psi_n^{(0)}(x). \tag{10.5}$$

Substitution of this in (10.4), multiplication of the resulting equation by $\psi_n^{(0)*}(x)$ on the left and integration over volume gives

$$(E_s - E_n^{(0)})c_{ns} = \sum_m \lambda c_{ms} V_{nm}, \tag{10.6}$$

where

$$V_{nm} = \int d^3x \psi_n^{(0)*}(x) V(x) \psi_m^{(0)}(x). \tag{10.7}$$

So far, no approximation has been made and (10.6) is exact. The first step towards obtaining approximate values of E_s and ψ_s is to expand c_{ns} and E_s in a power series in the small parameter λ:

$$\begin{aligned} c_{ns} &= c_{ns}^{(0)} + \lambda c_{ns}^{(1)} + \lambda^2 c_{ns}^{(2)} + \cdots \\ E_s &= E_s^{(0)} + \lambda E_s^{(1)} + \lambda^2 E_s^{(2)} + \cdots . \end{aligned} \tag{10.8}$$

Next, we substitute these series in (10.6) and equate coefficients of equal powers of λ on both sides of the resulting equation. This gives:

$$c_{ns}^{(0)} = \delta_{ns}, \qquad c_{ns}^{(1)} = \frac{V_{ns}}{E_s^{(0)} - E_n^{(0)}} \quad \text{for} \quad n \neq s. \tag{10.9}$$

$$E_n^{(1)} = V_{nn} = \int d^3x \psi_n^{(0)*}(x) V(x) \psi_n^{(0)}(x). \tag{10.10}$$

For finding $c_{nn}^{(1)}$, we normalise ψ_s retaining terms up to order λ:

$$\int d^3x (\psi_n^{(0)} + \lambda \psi_m^{(1)})^* (\psi_n^{(0)} + \lambda \psi_n^{(1)}) = 1 + O(\lambda^2) + \cdots \tag{10.11}$$

Neglecting terms of the order λ^2, we get from (10.11),

$$\int d^3x (\psi_n^{(0)*}(x) \psi_n^{(1)}(x) + \psi_n^{(1)*} \psi_n^{(0)}) = 0 \tag{10.12}$$

since

$$\int d^3x \psi_n^{(0)*}(x) \psi_n^{(0)}(x) = 1.$$

From (10.5), we have

$$\psi_n^{(1)}(x) = \sum_s c_{ns}^{(1)} \psi_s^{(0)}(x), \tag{10.13}$$

which on substitution in (10.12) gives

$$c_{nn}^{(1)} + c_{nn}^{(1)*} = 0.$$

Since the phase of the wave function is arbitrary, we take $c_{nn}^{(1)}$ to be real, in which case

$$c_{nn}^{(1)} = 0. \tag{10.14}$$

This, together with (10.9) determines $c_{ns}^{(1)}$ completely so that

$$\psi_n^{(1)}(x) = \sum_{s \neq n} \frac{V_{ns}\psi_s^{(0)}(x)}{E_n^{(0)} - E_s^{(0)}}, \tag{10.15}$$

where the sum is over the complete set and should include continuous eigenvalues over and above the discrete ones.

It may be noted here that if

$$V_{ns} > E_n^{(0)} - E_s^{(0)}, \tag{10.16}$$

then

$$\psi_n^{(1)}(x) > \psi_n^{(0)}(x) \tag{10.17}$$

for all values of x, in which case the series expansion (10.8) would not converge and the perturbation theory outlined above would fail.

Having obtained the first-order corrections to the energy eigenvalues and eigenfunctions, we now proceed to obtain the second-order corrections. For this, we substitute (10.8) in (10.6) and equate coefficients of λ^2 in the resulting equation. This gives

$$\left(E_s^{(0)} - E_n^{(0)}\right) c_{ns}^{(2)} + E_s^{(1)} c_{ns}^{(1)} + E_s^{(2)}\delta_{ns} = \sum_n c_{ms}^{(1)} V_{nm}. \tag{10.18}$$

Putting $n = s$ and on using (10.9) and (10.14), we obtain the second-order correction to the energy eigenvalue:

$$E_n^{(2)} = \sum_{m \neq n} \frac{|V_{nm}|^2}{E_n^{(0)} - E_m^{(0)}}. \tag{10.19}$$

If the unperturbed Hamiltonian has only two states labelled m and n, it follows from this equation that the higher level is pushed up and the lower level is pushed down as a result of second order perturbation. This is equivalent to a repulsion of energy levels. For $m \neq s$, (10.18) gives, on using (10.9) and (10.10),

$$c_{ns}^{(2)} = \sum_{m \neq s} \frac{V_{nm} V_{ms}}{(E_s^{(0)} - E_n^{(0)})(E_s^{(0)} - E_m^{(0)})} + \frac{V_{ss} V_{ns}}{(E_s^{(0)} - E_n^{(0)})^2}, \quad n \neq s. \tag{10.20}$$

To determine $c_{nn}^{(2)}$, we make use of the normalisation condition on $\psi(x)$ up to order λ^2 which gives

$$\int d^3x(\psi_n^{(2)*}(x)\psi_n^{(0)} + \psi_n^{(0)}(x)\psi_n^{(2)*} + \psi_n^{(1)*}\psi_n^{(1)}(x)) = 0, \tag{10.21}$$

which in terms of $c_{ns}^{(2)}$ and $c_{ns}^{(1)}$ becomes

$$c_{nn}^{(2)} + c_{nn}^{(2)*} + \sum_m |c_{nm}^{(1)}|^2 = 0. \tag{10.22}$$

As in the first-order case, we take $c_{nn}^{(2)}$ to be real and obtain

$$c_{nn}^{(2)} = -\frac{1}{2}\sum_m |c_{nm}^{(1)}|^2 = -\frac{1}{2}\sum_{m \neq n} \frac{|V_{nm}|^2}{(E_n^{(0)} - E_m^{(0)})^2}. \tag{10.23}$$

Substituting (10.20) and (10.23) in (10.5), we get for the second-order correction to the eigenfunctions:

$$\psi_n^{(2)} = \sum_{m \neq n} \sum_{s \neq n} \frac{V_{ms} V_{sn} \psi_m^{(0)}}{(E_n^{(0)} - E_s^{(0)})(E_n^{(0)} - E_m^{(0)})} - \sum_{m \neq n} \frac{V_{nn} V_{mn} \psi_n^{(0)}}{(E_n^{(0)} - E_m^{(0)})^2}$$
$$- \frac{1}{2} \sum_{m \neq n} \frac{|V_{nm}|^2 \psi_n^{(0)}}{(E_n^{(0)} - E_m^{(0)})^2}. \tag{10.24}$$

10.3 Time-Independent Perturbation for Degenerate States

If the quantum mechanical system has degenerate energy states, i.e., a number of unperturbed energy eigenvalues are equal, the terms in the sums in the formulae for corrections to these eigenvalues and their eigenfunctions given in the previous section will diverge on account of vanishing of energy denominators. This is because the method used therein has not taken degeneracy into account;it is meant for non-degenerate states. For degenerate states, a different method is needed.

For an unperturbed state which is degenerate, the Schrödinger equation can be written as

$$H_0 u_{rn} = E_n^{(0)} u_{rn}, \quad r = 1, 2, \cdots, m. \tag{10.25}$$

It is presumed that the eigenfunctions u_{rn} have been made orthonormal by the Schmidt method. Besides the m-fold degenerate state, the system may have many non-degenerate states, u_s. In that case the perturbed state n of the system can be represented as

$$\psi_n = \sum_{r=1}^{m} e_r u_{rn} + \sum_{s \neq n} c_{sn} u_s. \tag{10.26}$$

Substituting this in the perturbed Schrödinger equation

$$(H_0 + \lambda V)\psi_n = E_n \psi_n, \tag{10.27}$$

we get

$$\sum_{r=1} e_r E_n^{(0)} u_{rn} + \lambda \sum_{r=1} e_r V u_{rn} + \sum_{s \neq n} c_{sn} E_n^{(0)} u_s + \lambda \sum_{s \neq n} c_{sn} V u_s$$

$$= \sum_{r=1}^{n} e_r u_{rn} E_n + \sum_{s \neq n} c_{ns} E_n u_s. \tag{10.28}$$

Multiplying both sides of this equation on the left by u_{ln}^* and then integrating over all space we get

$$(E_n - E_n^{(0)})E_l = \lambda \sum_r e_r \langle ln|V|rn\rangle + \lambda \sum_{s \neq n} c_{sn} \langle ln|V|s\rangle \tag{10.29}$$

since u_{mn} are orthogonal to $u_{s \neq n}$. Further,

$$\psi_n \underset{\lambda=0}{\longrightarrow} \sum_r e_r u_{rn}, \tag{10.30}$$

$$c_{ns}^{(0)} = 0. \tag{10.31}$$

Let us now consider the first-order correction to degenerate state energy and eigenfunctions. To first-order λ (10.29) gives

$$E_n^{(1)} e_r = \sum_r e_r V_{lr}^{(n)},$$

where

$$V_{lr}^{(n)} = \langle ln|V|rn\rangle, \tag{10.32}$$

which can be written in matrix form as

$$\begin{pmatrix} V_{11}^{(n)} - E_n^{(1)} & V_{12}^{(n)} & V_{13}^{(n)} & \cdots & V_{1M}^{(n)} \\ V_{21}^{(n)} & V_{22}^{(n)} - E_n^{(1)} & V_{23}^{(n)} & \cdots & V_{2M}^{(n)} \\ \cdots & \cdots & \cdots & \cdots & \cdots \\ \cdots & \cdots & \cdots & \cdots & \cdots \\ \cdots & \cdots & \cdots & \cdots & \cdots \\ V_{M1}^{(n)} & V_{M2}^{(n)} & \cdots & \cdots & V_{MM}^{(n)} - E_n^{(1)} \end{pmatrix} \begin{pmatrix} e_1 \\ e_2 \\ \vdots \\ \vdots \\ \vdots \\ e_M \end{pmatrix} = 0. \tag{10.33}$$

The energy corrections are obtained by setting the determinant for the matrix on the left-hand side to zero. The corresponding eigenfunctions are obtained from this equation for each value of $E_n^{(1)}$.

Noting that $E_n^{(1)}$ is given by (10.33) and $c_{ns}^{(1)}$ is given correctly by the procedure for non-degenerate states, i.e.,

$$c_{ns}^{(1)} = \sum_{r \neq s} \frac{V_{sr}^{(n)} e_r^{(0)}}{E_r^{(0)} - E_s^{(0)}}, \tag{10.34}$$

the second-order correction is given by

$$\sum_r \left(\sum_s \frac{V_{ms} V_{sr}}{E_r^{(0)} - E_s^{(0)}} - E^{(2)} \delta_{mr} \right) e_r = 0. \tag{10.35}$$

Written out explicitly in matrix notation, the above has the form

$$\begin{pmatrix} \sum_s \left(\frac{V_{1s} V_{s1}}{E_1^{(0)} - E_s^{(0)}} \right) - E^{(2)} & \sum_s \frac{V_{1s} V_{s2}}{E_2^{(0)} - E_s^{(0)}} & \cdots \\ \sum_s \frac{V_{1s} V_{s1}}{E_1^{(0)} - E_s^{(0)}} & \sum_s \left(\frac{V_{2s} V_{s2}}{E_2^{(0)} - E_s^{(0)}} \right) - E^{(2)} & \cdots \\ \cdots & \cdots & \cdots \end{pmatrix} \begin{pmatrix} e_1 \\ e_2 \\ \vdots \end{pmatrix} \tag{10.36}$$

whose solution is obtained as in the first-order case.

We shall apply the above theory to problems of linear Stark effect in hydrogen and charged particle moving on a circle in two dimensions.

Linear Stark effect in hydrogen

We consider the simple case of $n = 2$ state of hydrogen atom whose four states with eigenfunction u_{211}, u_{21-1}, u_{210} and u_{200} have the same energy. The perturbation caused by an external uniform electric field $\mathcal{E}$, is

$$V = -ez\mathcal{E}. \tag{10.37}$$

Its only non-zero matrix elements are

$$\langle u_{210}|V|u_{200}\rangle = \langle u_{200}|V|u_{210}\rangle = 3ea_0\mathcal{E} \tag{10.38}$$

so that matrix equation (10.33) takes the form

$$\begin{pmatrix} -E_2^{(1)} & 3ea_0\mathcal{E} \\ 3ea_0\mathcal{E} & -E_2^{(1)} \end{pmatrix} \begin{pmatrix} e_1 \\ e_2 \end{pmatrix} = 0. \tag{10.39}$$

The solution of this equation is

$$E_2^{(1)} = \pm 3ea_0\mathcal{E}. \tag{10.40}$$

For $E_2^{(1)} = +3ea_0\mathcal{E}$, $e_1 = e_2$ so that the normalised wavefunction is

$$\psi_2^{(+)} = \frac{1}{\sqrt{2}}(u_{210} + u_{200}) \tag{10.41}$$

and for $E_2^{(1)} = -3ea_0\mathcal{E}$, $e_1 = -e_2$ and the corresponding wavefunction is

$$\psi_2^{(-)} = \frac{1}{\sqrt{2}}(u_{200} - u_{210}). \tag{10.42}$$

The energies of these two states to first order in $\mathcal{E}$ are

$$E_2^{(\pm)} = E_2^{(0)} + E_2^{(1)}. \tag{10.43}$$

Stark beats

In general, the wave function will be a linear superposition of $\psi_2^{(+)}$ and $\psi_2^{(-)}$: Dropping the subscript 2, we have

$$\psi = a\psi^{(+)} + b\psi^{(-)}, \tag{10.44}$$

whose time dependence can be written explicitly as

$$\psi(t) = \frac{a}{\sqrt{2}}(u_{200} + u_{210})e^{-\frac{i}{\hbar}E^{(1)}t} + \frac{b}{\sqrt{2}}(u_{200} - u_{210})e^{-\frac{i}{\hbar}E^{(-)}t}. \tag{10.45}$$

The probability of finding the state $|200\rangle$ in this is

$$P_{200}(t) = |\langle u_{200}|\psi(t)\rangle|^2 = \frac{a^2 + b^2}{2} + ab\cos\frac{\Delta Et}{\hbar} \tag{10.46}$$

and that of the state $|210\rangle$ is

$$P_{210}(t) = \frac{a^2 + b^2}{2} - ab\cos\frac{\Delta Et}{\hbar}, \tag{10.47}$$

where

$$\Delta E = E^{(+)} - E^{(-)} = 6ea_0\mathcal{E}. \tag{10.48}$$

Since ψ in (10.44) has to be normalised to unity, we take $|a|^2 = |b|^2 = \frac{1}{2}$. If initially ($t = 0$) the atom is in the state $|200\rangle$, $a = b$ and

$$P_{200}(t) = \frac{1}{2}\left(1 + \cos\frac{\Delta E t}{\hbar}\right)$$

$$P_{210}(t) = \frac{1}{2}\left(1 - \cos\frac{\Delta E t}{\hbar}\right). \tag{10.49}$$

If on the other hand initially the atom is in the state $|210\rangle$, $b = -a$ and

$$P_{200}(t) = \frac{1}{2}\left(1 - \cos\frac{\Delta E t}{\hbar}\right)$$

$$P_{210}(t) = \frac{1}{2}\left(1 + \cos\frac{\Delta E t}{\hbar}\right). \tag{10.50}$$

In both cases $P_{200}(t) + P_{210}(t) = 1$, as it ought to be. The probabilities oscillate with time; this is known as Stark oscillation or beat. These oscillations show up in the intensities of Lyman α lines and have been experimentally observed. Similar oscillations occur when the atom is in an external uniform magnetic field; these are known as Zeeman beats and have also been observed. Stark and Zeeman beats are examples of quantum beats which occur due to mixing of close-by neighbouring states caused by their interaction. If an atom is excited by appropriate laser beams to two neighbouring excited states, the intensity of radiation from these states will show oscillatory behaviour.

Linear Stark effect in plane rotator*

We consider the case of a charged particle confined to a circle of radius a, on which a uniform external electric field is imposed. The Hamiltonian of the system is given by

$$H = H_0 + V \tag{10.51}$$

where

$$H_0 = -\frac{p_\varphi^2}{8ma^2}, \quad V = e\mathcal{E}a(1 - \cos 2\varphi). \tag{10.52}$$

*In many books this problem has been solved using non-degenerate perturbation theory disregarding the double degeneracy of each level except the ground state, which is non-degenerate.

The unperturbed problem has been solved in chapter 7, section 4. Instead of the angle θ, we have here $\varphi = \theta/2$. The unperturbed eigenvalues and eigenfunctions are

$$E_n^{(0)} = \frac{n^2\hbar^2}{2ma^2} \qquad n = 0, 1, 2 \cdots, \tag{10.53a}$$

$$\psi_n^{(0)} = \frac{1}{\sqrt{\pi}}\, e^{\pm 2in\varphi}. \tag{10.53b}$$

It is to be noted that all levels except $n = 0$ are doubly degenerate. For the level n, the first-order correction $E_n^{(1)}$ to the energy eigenvalue is obtained from the determinantal equation

$$\begin{vmatrix} V_{n_+,n_+} - E^{(1)} & V_{n_+,n_-} \\ V_{n_-,n_+} & V_{n_-,n_-} - E^{(1)} \end{vmatrix} = 0, \tag{10.54}$$

where

$$V_{n_+,n_+} = \langle n_+|V|n_+\rangle, \quad \text{etc.,}$$

$$n_\pm = \frac{1}{\sqrt{\pi}} e^{\pm 2in\varphi}. \tag{10.55}$$

Using (10.52) and (10.53), we get

$$\begin{aligned} V_{n_+n_+} &= \frac{e\mathcal{E}a}{\pi}\int_0^\pi d\varphi e^{-2in\phi}(1 - \cos 2\varphi)e^{2in\varphi} \\ &= e\mathcal{E}a, \end{aligned} \tag{10.56}$$

and similarly,

$$V_{n_-n_-} = e\mathcal{E}a \ , \ \ V_{n_+n_-} = V_{n_-n_+} = 0\,. \tag{10.57}$$

Substituting (10.56) and (10.57) in (10.54) gives

$$E^{(1)} = e\mathcal{E}a, \tag{10.58}$$

which shows that the corrections to both $|n+\rangle$ and $|n-\rangle$ are equal. Thus the first-order perturbation does not remove the two-fold degeneracy.

The second-order correction is obtained from the determinantal equation

$$\begin{vmatrix} \sum_{n'}' \frac{V_{n_+n'}V_{n'n_+}}{E_n^{(0)} - E_{n'}^{(0)}} - E^{(2)} & \sum_{n'}' \frac{V_{n_+n'}V_{n'n_-}}{E_n^{(0)} - E_{n'}^{(0)}} \\ \sum_{n'}' \frac{V_{n_-n'}V_{n'n_+}}{E_n^{(0)} - E_{n'}^{(0)}} & \sum_{n'}' \frac{V_{n_-n'}V_{n'n_-}}{E_n^{(0)} - E_{n'}^{(0)}} - E^{(2)} \end{vmatrix} = 0. \tag{10.59}$$

The primes on the summation sign indicate that the states $|n_\pm\rangle$ are not to be included in the sums. In general, $V_{nn'}$ works out to

$$V_{nn'} = \langle n|V|n'\rangle = \frac{ea\mathcal{E}}{\pi}\int_0^\pi d\varphi e^{2i(n-n')\varphi}(1-\cos 2\varphi)$$
$$= e\mathcal{E}a\left(\delta_{nn'} - \frac{1}{2}\delta_{n,n'\pm 1}\right). \tag{10.60}$$

Since the states $n = n'$ are to be excluded,

$$V_{nn'} = -\frac{1}{2}e\mathcal{E}a\delta_{n,n'\pm 1}. \tag{10.61}$$

For the state $n = 1$, while $n' = n-1 = 0$ is non-degenerate, $n' = n+1 = 2$ is degenerate whereas for $n > 1$ both the $n' = n-1$ and $n' = n+2$ are degenerate. Therefore the state $n = 1$ requires special treatment. For $n > 1$, (10.59) becomes, in view of (10.61),

$$\begin{vmatrix} \dfrac{|V_{n_+,n_++1}|^2}{E_n^{(0)} - E_{n+1}^{(0)}} + \dfrac{|V_{n+,n_+-1}|^2}{E_n^{(0)} - E_{n-1}^{(0)}} - E^{(2)} & 0 \\ 0 & \dfrac{|V_{n_-,n_-+1}|^2}{E_n^{(0)} - E_{n+1}^{(0)}} + \dfrac{|V_{n_-,n_--1}|^2}{E_n^{(0)} - E_{n-1}^{(0)}} - E^{(2)} \end{vmatrix}. \tag{10.62}$$

The use of (10.61) for the matrix elements in this equation gives

$$E_n^{(2)} = \frac{(ma^2)^3(e\mathcal{E}/am)^2}{(4n^2-1)\hbar^2}; \tag{10.63}$$

the two-fold degeneracy still persists. For $n = 1$ we have

$$\begin{vmatrix} \dfrac{e^2m\mathcal{E}^2a^4}{3\hbar^2} - E^{(2)} & \dfrac{e^2m\mathcal{E}^2a^4}{3\hbar^2} \\ \dfrac{e^2m\mathcal{E}^2a^4}{2\hbar^2} & \dfrac{e^2m\mathcal{E}^2a^4}{3\hbar^2} - E^{(2)} \end{vmatrix} = 0, \tag{10.64}$$

where the two solutions are

$$E^{(2)} = \frac{1}{6\hbar^2}\begin{cases} -(ma^2)^3\left(\frac{\mathcal{E}e}{a_m}\right)^2, \\ 5(ma^2)^3\left(\frac{e\mathcal{E}}{a_m}\right)^2. \end{cases} \tag{10.65}$$

Thus the degeneracy is removed in second order. On calculation, one finds that in the third-order of perturbation, degeneracy will be removed only for $n = 2$

and in general degeneracy of a state $|n\rangle$ will be removed in the $(n+1)$th and higher orders.

For the state $|n=0\rangle$, which is non-degenerate,

$$E^{(1)}_{n=0} = \langle n=0|V|n=0\rangle = e\mathcal{E}a,$$

$$E^{(2)}_{n=0} = \sum_{n'}{}' \frac{V_{0n'}V_{n'0}}{E^{(0)}_0 - E^{(0)}_{n'}} = -\frac{(ma^2)^3\left(\frac{e\mathcal{E}}{a}\right)^2}{\hbar^2}. \tag{10.66}$$

10.4 Time-Dependent Perturbation

Schrödinger picture

So far, we have seen how time-independent perturbation brings about changes in the eigenvalues and eigenfunctions of unperturbed systems. For perturbation explicitly dependent on time, the total Hamiltonian being time-dependent, there is lack of conservation of energy and there are no stationary states. As such, corrections to unperturbed eigenvalues and eigenfunctions are not meaningful. Such perturbations cause quantum transitions between unperturbed states of the system. To deal with such transitions, we have to solve the time-dependent Schrödinger equations

$$i\frac{\partial\psi_n}{\partial t} = (H_0 + V(t))\psi_n, \tag{10.67}$$

$$i\frac{\partial\psi^{(0)}_n}{\partial t} = H_0\psi^{(0)}_n. \tag{10.68}$$

For this, we expand $\psi_n(t)$ in terms of $\psi^{(0)}_n$:

$$\psi_n(t) = \sum_k a_{kn}(t)\psi^{(0)}_k \tag{10.69}$$

and substitute it in (10.67) to get

$$i\frac{da_{mn}}{dt} = \sum_k V_{mk}(t)a_{km}(t), \tag{10.70}$$

where

$$V_{mk}(t) = \int d^3r\,\psi^{(0)*}_m(\vec{r}t)V(\vec{r}t)\psi^{(0)}_k(\vec{r}t). \tag{10.71}$$

The exact solution of (10.70) is not possible except for some simple model systems. For the general case, we have to obtain approximate solutions. As a first approximation we shall take

$$a_{kn} = a_{kn}^{(0)} + a_{kn}^{(1)} \quad , \quad a_{kn}^{(1)} \ll a_{kn}^{(0)}. \tag{10.72}$$

Let the system be in the nth unperturbed state at $t = 0$ to which the perturbation is applied. In this case,

$$a_{kn}(t = 0) = a_{kn}^{(0)} = \delta_{kn}. \tag{10.73}$$

Substituting (10.72) in (10.70) and making use of the initial condition (10.73), we get

$$i\frac{da_{kn}^{(1)}}{dt} = V_{kn}(t), \tag{10.74}$$

which on integration gives

$$\begin{aligned} a_{kn}^{(1)}(t) &= -i\int dt V_{kn}(t) = -i\int dt \int d^3r \psi_k^{(0)*}(\vec{r}t)V(\vec{r}t)\psi_n^{(0)}(\vec{r}t) \\ &= -i\int dt \int d^3r u_k^*(\vec{r})V(\vec{r}t)u_n(\vec{r})e^{i\omega_{kn}t}, \end{aligned} \tag{10.75}$$

where

$$u_n(\vec{r}) = e^{iE_n t}\psi_n^{(0)}(\vec{r}t), \tag{10.76}$$

$$V_{kn} = \int d^3r u_k^* V_{u_n} e^{i\omega_{kn}t}, \tag{10.77}$$

$$\omega_{kn} = E_k - E_n, \tag{10.78}$$

E_n being the energy eigenvalue of the unperturbed Hamiltonian.

If the perturbation is switched on adiabatically to a constant operator F,

$$V(t) = e^{\lambda t}F, \tag{10.79}$$

where λ is an infinitesimally small parameter such that $V(-\infty) = 0$; then (10.75) leads to

$$a_{kn}^{(1)} = -F_{kn}\frac{e^{i\omega_{kn}t+\lambda t}}{(\omega_{kn} - i\lambda)}. \tag{10.80}$$

The transition probability per unit time W_{kn} defined by

$$W_{kn} = \frac{d}{dt}|a_{kn}^{(1)}|^2 \tag{10.81}$$

obtained from (10.80) comes out to be

$$W_{kn} = 2\pi e^{2\lambda t}|F_{kn}|^2\frac{\lambda}{(\omega_{kn}^2 + \lambda^2)\pi}, \tag{10.82}$$

where the limit $\lambda \to 0$ is to be taken. Since

$$\lim_{\lambda \to 0} \frac{\lambda}{\pi(\lambda^2 + \alpha^2)} = \delta(\alpha), \tag{10.83}$$

the limit $\lambda \to 0$ of (10.81) turns out to be

$$W_{kn} = 2\pi |F_{kn}|^2 \delta(E_k - E_n). \tag{10.84}$$

For transition to a continuum of final states with density of states $\rho(E_f)$ from a discrete state,

$$W_{fi} = 2\pi |F_{fi}|^2 \int dE_f \rho(E_f) \delta(E_i - E_f)$$

$$= 2\pi |F_{fi}|^2 \rho(E_i) \tag{10.85}$$

which is known as 'Fermi Golden Rule'.

If, instead of a constant potential as treated above F is time-dependent, i.e., there is a periodic potential, then,

$$F(t) = f e^{i\omega t} + f^+ e^{-i\omega t} \tag{10.86}$$

$$a_{kn}^{(1)} = -f_{kn} \frac{e^{i(\omega_{kn} - \omega)t}}{\hbar(\omega_{kn} - \omega)} - f_{kn}^* \frac{e^{i(\omega_{kn} + \omega)t}}{\hbar(\omega_{kn} + \omega)}. \tag{10.87}$$

Dirac picture

Since the potential is time-dependent, the problem can be better handled if we adopt the Dirac picture of quantum mechanics instead of the Schrödinger picture presented above (where only the lowest-order calculation is given). As we have seen in chapter 2, the operators and the wavefunction in the Dirac picture are time-dependent in contrast to those of Schrödinger picture where the operators are time-independent and the wavefunctions carry the time-dependence. The Dirac picture is therefore better suited to time-dependent perturbation. It will be recalled (chapter 2) that the wavefunctions and operators of the Dirac picture are related to those of the Schrödinger picture by the unitary transformations

$$\psi_D(t) = e^{iH_0 t} \psi_s(t),$$

$$\Omega_D(t) = e^{iH_0 t} \Omega_s e^{-iH_0 t}, \tag{10.88}$$

so that the equations of motion are

$$i \frac{\partial \psi_D(t)}{\partial t} = \lambda V_D \psi_D(t),$$

$$\frac{d\Omega_D(t)}{dt} = \frac{\partial \Omega_D}{\partial t} + i[H_0, \Omega_D], \tag{10.89}$$

where

$$V_D(t) = e^{iH_0t}V_s e^{-iH_0t}.$$

The integral version of the first equation is given by

$$\psi_D(t) = \psi_D^{(0)}(t) - i\lambda \int_{-\infty}^{t} dt' V_D(t')\psi_D(t') \tag{10.90}$$

whose iterated solution is

$$\psi_D(t) = \psi_D^{(0)}(t) - i\lambda \int_{-\infty}^{t} dt' V_D(t')\psi_D^{(0)}(t) + (i\lambda)^2 \int_{-\infty}^{t} dt' V_D(t') \times \int_{-\infty}^{t'} dt'' V_d(t'')\psi_D^{(0)}(t'') + \cdots . \tag{10.91}$$

This can be written in terms of Dyson's P-product as

$$\psi_D(t) = \left[1 + (-i\lambda) \int_{-\infty}^{t} dt' V_D(t') + \left(\frac{-i\lambda}{2!}\right)^2 \int_{-\infty}^{t} dt' \times \int_{-\infty}^{\infty} dt'' P(V_d(t')V_D(t'')) + \cdots \right] \psi_D^{(0)}, \tag{10.92}$$

where

$$P(f(t_1)g(t_2)) = \begin{cases} f(t_1)g(t_2) & \text{for } t_1 > t_2 \\ g(t_2)f(t_1) & \text{for } t_2 > t_1 \end{cases} \tag{10.93}$$

since $\psi_D^{(0)}$ is time-independent as follows from (10.89). In closed form (10.92) can be written as

$$\psi_D(t) = P \exp\left[-i \int_{-\infty}^{t} dt' V_D(t')\right] \psi_D^{(0)}. \tag{10.94}$$

For a system in a state $|n\rangle$, (10.90) can be written as

$$\psi_D(t) = \psi_n(t) = \phi_n - i\lambda \int_{-\infty}^{t} dt' V(t')\psi_n(t'), \tag{10.95}$$

where ϕ_n is the unperturbed stationary state. Expanding ψ_n in a complete set of unperturbed states

$$\psi_n(t) + \sum_s a_{sn}(t)\phi_s, \tag{10.96}$$

we get

$$a_{sn} = \delta_{sn} - i\lambda \int_{-\infty}^{t} dt' \langle s|V(t')|\psi_n(t)\rangle$$

where $\psi_n(t)$ is given by (10.94) in terms of ϕ_n.

Problems

1. A hydrogen atom is perturbed by a potential

$$V = +\frac{a}{r^2}.$$

Calculate first-order correction to the energy eigenvalues of 1s, 2s and 2p levels and compare with exact theory.
Note: The exact solution is given in *Quantum Mechanics* by Landau and Lifshitz (3rd edition, problem 3, chapter 36).

2. Calculate the first-order Stark splittings for the hydrogen atom in the state $n = 3$. Find the eigenfunctions of the split levels.
Hint: Number the degenerate states so that the perturbation eigenvalue matrix has 2×2 and 3×3 matrices along the diagonal.

3. A particle constrained to move on the surface of a sphere of radius a is perturbed by a potential

$$V = v\cos^2\theta.$$

Obtain the first corrections to energy eigenvalues for states with $l = 2$ using degenerate perturbation theory.

4. A beam of atoms with spin $\frac{1}{2}$ are made to pass through a Stern–Gerlach field which splits it into spin-up and spin-down beams. The spin-up beam is then made to pass through a combination of homogeneous constant and alternating magnetic fields at right angles to each other (the former along z-axis and the latter along x-axis). The emergent beam is then analysed by a second Stern–Gerlach field. Find the ratio of number of atoms with spin-up and spin-down in the final beam using perturbation theory.
Hint: Apply time-dependent perturbation theory with

$$H_0 = -\mu B\sigma_z, \quad H_I = -\mu a\sigma_x$$

to obtain transition amplitude from the state with $\psi_i = \begin{pmatrix}1\\0\end{pmatrix}$ and $\psi_f = \begin{pmatrix}0\\1\end{pmatrix}$.

Bibliography

[1] H.J. Andra, *Phys. Rev.*, A**2**, 2200 (1970).
[2] T. Pradhan and A.V. Khare, *Am. J. Phys.*, **41**, 59 (1973).

11 Electron Spin and Hydrogen Fine Structure

11.1 Electron Spin

The notion of electron having an intrinsic angular momentum (spin) of $\frac{1}{2}\hbar$ was put forth by Uhlenbeck and Goudsmit to explain the observed fine structure of line spectra of atoms and the results of Stern–Gerlach experiment. Unlike orbital angular momentum discussed in Chapter 4, which is determined by the nature of motion of the electron in configuration space and has a classical limit, the spin angular momentum is independent of its orbital motion and disappears in this limit ($\hbar \to 0$). Calculation of additional energy of the atom arising from electron spin was made by Pauli, who developed the quantum mechanical theory for it in the non-relativistic framework. Later, Dirac gave a relativistic theory of the electron in which spin arises naturally and for which Pauli theory emerges in the non-relativistic limit.

For a particle with spin, the wavefunction ψ is characterised by a discrete variable in addition to its space coordinates. This variable is a projection of its spin angular momentum along the z-axis. Thus,

$$\psi \equiv \psi(\vec{r}, s), \tag{11.1}$$

where s for spin $\frac{1}{2}$ takes the values $+\frac{1}{2}$ and $-\frac{1}{2}$. The square modulus of this wavefunction integrated over space gives the probability $P(s)$ that the particle has a certain spin projection value s:

$$P(s) = \int d^3r |\psi(\vec{r}, s)|^2. \tag{11.2}$$

Similarly, the same square modulus summed over all discrete spin projections s gives the probability $P(\vec{r})$ of finding the particle at $\vec{r}$:

$$P(\vec{r}) = \sum_s |\psi(\vec{r}, s)|^2. \tag{11.3}$$

The spin wavefunction $\xi(s)$ can be defined by

$$P(s) = |\xi(s)|^2. \tag{11.4}$$

For the electron which has spin $\frac{1}{2}$, the wavefunction $\xi(s)$ has two components:

$$\xi(s) = \begin{pmatrix} \xi_1 \\ \xi_2 \end{pmatrix}, \tag{11.5}$$

where ξ_1 and ξ_2 are in general, complex numbers. The angular momentum operators S_i in this space are

$$S_i = \frac{1}{2}\sigma_i,$$

$$\sigma_1 = \begin{pmatrix} 0 & 1 \\ 1 & 0 \end{pmatrix}, \quad \sigma_2 = \begin{pmatrix} 0 & -i \\ i & 0 \end{pmatrix}, \quad \sigma_3 = \begin{pmatrix} 1 & 0 \\ 0 & -1 \end{pmatrix}, \tag{11.6}$$

introduced by Pauli in his theory of electron spin. Rotations in this space are effected by three families of 2×2 unitary ($UU^+ = U^+U = 1$), unimodular (determinant $= 1$) matrices which form a $SU(2)$ group:

$$U_i = e^{\frac{i}{2}\sigma_i\theta_i}, \text{ (no summation)} \tag{11.7}$$

where θ_i are parameters of transformation. Such an $SU(2)$ transformation on ξ_α is equivalent to a rotation in corresponding three-dimensional space with

$$r_i = \frac{1}{2}\,\xi^\dagger\sigma_i\xi \tag{11.8}$$

effected by

$$R_i = e^{iS_i\theta_i}, \tag{11.9}$$

where

$$S_1 = \begin{pmatrix} 0 & 0 & 0 \\ 0 & 0 & -i \\ 0 & i & 0 \end{pmatrix}, \quad S_2 = \begin{pmatrix} 0 & 0 & i \\ 0 & 0 & 0 \\ -i & 0 & 0 \end{pmatrix}, \quad S_3 = \begin{pmatrix} 0 & -i & 0 \\ i & 0 & 0 \\ 0 & 0 & 0 \end{pmatrix}, \tag{11.10}$$

and θ_i are the angles of rotation about the ith axis. In order to show this, we start with the infinitesimal transformation

$$\delta\xi_\alpha = \frac{i\theta_i}{2}\sigma_i^{\alpha\beta}\xi_\beta \tag{11.11}$$

corresponding to (11.7) and use it in

$$\delta r_i = \frac{i}{2}(\delta\xi_\alpha^+ \sigma_i^{\alpha\beta}\xi_\beta + \xi_\alpha^+ \sigma_i^{\alpha\beta}\delta\xi_\beta) \tag{11.12}$$

obtained from (11.8). After few algebraic steps, we get

$$\delta r_i = i\mathcal{E}_{ijk}\theta_j \xi^+ \sigma_k \xi = i\theta_j (S_j)_{ik} r_k, \tag{11.13}$$

which is the infinitesimal form of (11.9).

The spin wavefunction ξ can be expressed in terms of spherical polar coordinates

$$x = \sin\theta\cos\phi, \quad y = \sin\theta\sin\phi, \quad z = \cos\theta, \tag{11.14}$$

by making use of (11.8). Simple algebraic manipulation gives

$$\xi = \begin{pmatrix} \cos\theta/2 & e^{-i\phi/2} \\ \sin\theta/2 & e^{+i\phi/2} \end{pmatrix} \tag{11.15}$$

apart from an overall phase factor.

11.2 Hydrogen Fine Structure (Pauli Theory)

The spectral lines of hydrogen atom are not single but exhibit a fine structure. This is attributed to the spin of the electron. The spin magnetic moment interacts with the magnetic field that the electron moving in a Coulomb field of the nucleus experiences. This is called spin–orbit interaction, which together with relativistic corrections gives the correct splitting of energy levels and fine structure of spectral lines.

The magnetic field $\vec{B}$ that the electron experiences due to its motion in the Coulomb field $\vec{E} = e\vec{r}/r^3$ of the nucleus is given by

$$\vec{B} = \vec{E} \times \vec{v}/c = \frac{e\hbar}{mc}\frac{\vec{L}}{r^3}, \tag{11.16}$$

and the magnetic moment associated with the spin motion of the electron is

$$\vec{\mu}_s = \frac{e\vec{\sigma}}{2mc}. \tag{11.17}$$

The interaction energy is therefore

$$H_{LS} = \vec{\mu}_s \cdot \vec{B} = \frac{-e^2}{2m^2c^2}\frac{1}{r^3}\vec{L}\cdot\vec{\sigma}. \tag{11.18}$$

This is off by a factor of $\frac{1}{2}$ from relativistic treatment, which gives $\vec{L} \cdot \vec{S}$ instead of $\vec{L} \cdot \vec{\sigma}$ in the above formula. The relativistic correction is obtained by going to the limit $p^2/m^2c^2 \ll 1$ of the relativistic kinetic energy

$$E = (p^2c^2 + m^2c^4)^{1/2} = mc^2\left(1 + \frac{p^2}{m^2c^2}\right)^{1/2}$$

$$= mc^2\left(1 + \frac{p^2}{2m^2c^2} - \frac{p^4}{8m^4c^4} + \cdots\right)$$

$$= mc^2 + \frac{p^2}{2m} - \frac{p^4}{8m^3c^2} + \cdots. \tag{11.19}$$

Apart from (11.19), there is another relativistic correction term called the Darwin term

$$H_{\text{Darwin}} = \frac{\pi e^2}{2m^2c^2}\delta(\vec{r}). \tag{11.20}$$

The total perturbation due to the spin–orbit interaction and relativity correction is thus (taking $c = 1$)

$$H_I = -\frac{p^4}{8m^3} + \frac{\alpha}{2m^2}\frac{\vec{L} \cdot \vec{S}}{r^3} + \frac{\alpha\pi}{2m^2}\delta(\vec{r}) \tag{11.21}$$

with

$$H_0 = \frac{p^2}{2m} - \frac{e^2}{r}.$$

$H = H_0 + H_I$ is the Pauli Hamiltonian.

In order to obtain the first-order correction to energy, we have to take the expectation value of each term of (11.21) for which we note that

$$p^2\psi_{nl} = 2m\left(E_n^{(0)} + \frac{\alpha}{r}\right)\psi_{nl}$$

$$p^4\psi_{nl} = 2m\left(E_n^{(0)} + \frac{\alpha}{r}\right)p^2\psi_{nl} = 4m^2\left(E_n^{(0)} + \frac{\alpha}{r}\right)^2\psi_{nl}$$

$$\langle p^4\rangle_{nl} = 4m^2\left\langle\left(E_n^{(0)} + \frac{\alpha}{r}\right)^2\right\rangle_{nl}$$

$$= 4m^2(E_n^{(0)})^2 + \alpha^2\left\langle\frac{1}{r^2}\right\rangle_{nl} + 2E_{en}^{(0)}\left\langle\frac{1}{r}\right\rangle_{nl}. \tag{11.22}$$

On calculation, one finds that (taking $\hbar = 1$)

$$\left\langle \frac{1}{r} \right\rangle_{nl} = \frac{m\alpha}{n^2},$$

$$\left\langle \frac{1}{r^2} \right\rangle_{nl} = \frac{m^2\alpha^2}{n^2(l+\frac{1}{2})}. \tag{11.23}$$

Substituting (11.23) in (11.22) gives

$$\langle p^4 \rangle_{nl} = -\frac{m\alpha^4}{n^4}\left(\frac{n}{l+\frac{1}{2}} - \frac{3}{4}\right). \tag{11.24}$$

Further,

$$\langle \vec{L} \cdot \vec{S} \rangle_{nl} \begin{cases} \frac{1}{2}\left[j(j+1) - l(l+1) - \frac{3}{4}\right] & \text{for } l \neq 0 \\ 0 & \text{for } l = 0 \end{cases} \tag{11.25}$$

where $j = l \pm \frac{1}{2}$,

$$\left\langle \frac{1}{r^3} \right\rangle_{nl} = \frac{m^3\alpha^3}{n^3 l(l+1)\left(l+\frac{1}{2}\right)} \tag{11.26}$$

$$\langle \delta(\vec{r}) \rangle_{nl} = |\psi_{nl}(r=0)|^2 = \begin{cases} \frac{1}{\pi}\left(\frac{\alpha m}{n}\right)^3 & \text{for } l = 0 \\ 0 & \text{for } l \neq 0. \end{cases} \tag{11.27}$$

Making use of (11.24) to (11.27), the expectation value of H_I in (11.21) adds up to

$$\Delta E_{nl} = \langle H_I \rangle_{nl} = -\frac{m\alpha^4}{2n^3}\left(\frac{1}{j+\frac{1}{2}} - \frac{3}{4n}\right). \tag{11.28}$$

It is to be noted that the correction depends on j alone. Thus the l degeneracy is not completely removed. For instance, the pairs $2^2S_{1/2}$ and $2^2P_{1/2}$, $3^2S_{1/2}$, and $3^2P_{1/2}$, $3^2P_{3/2}$ and $3^2D_{3/2}$ are still degenerate while $2^2P_{3/2}$ and $3^2D_{5/2}$ are non-degenerate. The level with quantum number n splits into n fine structure components.

11.3 Pauli Eigenfunction

On account of the spin–orbit interaction term, the eigenfunction will be a two component spinor, which in general will have the form

$$u = \begin{pmatrix} u_1 \\ u_2 \end{pmatrix} = R_{nl} \begin{pmatrix} aY_l, m_l(\theta, \varphi) \\ bY_l, m_l'(\theta, \varphi) \end{pmatrix}. \tag{11.29}$$

From this eigenfunction we have

$$J_3 \begin{pmatrix} u_1 \\ u_2 \end{pmatrix} = (L_3 + S_3) \begin{pmatrix} u_1 \\ u_2 \end{pmatrix} = \begin{pmatrix} \left(m_l + \frac{1}{2}\right) u_1 \\ \left(m_l' - \frac{1}{2}\right) u_2 \end{pmatrix}. \tag{11.30}$$

The eigenvalue equation

$$J_3 \begin{pmatrix} u_1 \\ u_2 \end{pmatrix} = m_j \begin{pmatrix} u_1 \\ u_2 \end{pmatrix} \tag{11.31}$$

coupled with (11.30) gives

$$m_l = m_j - \frac{1}{2}, \quad m_l' = m_j + \frac{1}{2}.$$

Substituting these values in (11.29) and operating on it by $\vec{L} \cdot \vec{S}$ gives

$$\begin{aligned} 2\vec{L} \cdot \vec{S} = \begin{pmatrix} u_1 \\ u_2 \end{pmatrix} &= \begin{pmatrix} L_3 & L_- \\ L_+ & L_3 \end{pmatrix} \begin{pmatrix} u_1 \\ u_2 \end{pmatrix} \\ &= R_{nl}(r) \begin{pmatrix} \left[\left(m_j - \frac{1}{2}\right) a - b\sqrt{\left(l + \frac{1}{2}\right)^2 - m_j^2}\right] Y_{l,m_j - \frac{1}{2}} \\ \left[-a\sqrt{\left(l + \frac{1}{2}\right)^2 - m_j^2} - b\left(m_j + \frac{1}{2}\right)\right] Y_{l,m_j + \frac{1}{2}} \end{pmatrix}. \end{aligned} \tag{11.32}$$

The eigenvalue equation

$$2(\vec{L} \cdot \vec{S}) \begin{pmatrix} u_1 \\ u_2 \end{pmatrix} = \lambda \begin{pmatrix} u_1 \\ u_2 \end{pmatrix} \tag{11.33}$$

is then satisfied for two values for a/b:

$$\left(\frac{a}{b}\right)_+ = -\sqrt{\frac{l + m_j + \frac{1}{2}}{l - m_j + \frac{1}{2}}}, \quad \left(\frac{a}{b}\right)_- = \sqrt{\frac{l - m_j + \frac{1}{2}}{l + m_j + \frac{1}{2}}} \tag{11.34}$$

and the eigenvalues for these are

$$\lambda_+ = l \;\; \lambda_- = -(l + 1) \tag{11.35}$$

except for $l = 0$, for which $\lambda_+ = \lambda_- = 0$. The corresponding normalised eigenfunctions then take the form

$$u_{n,l,j=l+\frac{1}{2},m_j}(r,\theta,\varphi) = \frac{1}{\sqrt{2l+1}} R_{nl}(r) \begin{pmatrix} \sqrt{l+m_j+\frac{1}{2}} & Y_{l,m_j-\frac{1}{2}}(\theta,\varphi) \\ -\sqrt{l-m_j+\frac{1}{2}} & Y_{l,m_j+\frac{1}{2}}(\theta,\varphi) \end{pmatrix}, \tag{11.36}$$

$$u_{n,l,j=l-\frac{1}{2},m_j}(r,\theta,\varphi) = \frac{1}{\sqrt{2l+1}} R_{nl}(r) \begin{pmatrix} \sqrt{l-m_j+\frac{1}{2}} & Y_{l,m_j-\frac{1}{2}}(\theta,\varphi) \\ \sqrt{l+m_j+\frac{1}{2}} & Y_{l,m_j+\frac{1}{2}}(\theta,\varphi) \end{pmatrix}, \tag{11.37}$$

the angular parts of which are the same as $\Omega_{lm_l}(\theta,\varphi)$ and $\Omega_{(-l+1),m_l}(\theta,\varphi)$ of equations (9.8a) and (9.8b) of chapter 9 except for the phase factor.

Problems

1. Obtain an expression for the permanent electric dipole moment of the hydrogen atom with electron spin for a state n, j and m_j.
   Hint: See section 55 of reference [3] below.

Bibliography

[1] W. Pauli, *Z. Physik*, **43**, 601 (1927).

[2] L.H. Thomas, *Nature* (London), **107**, 514 (1926).

[3] H.A. Bethe and E.E. Salpeter, Quantum Mechanics of One and Two Electron System, in *Encyclopedia of Physics*, Vol. XXXV, Springer Verlag, Berlin (1957).

[4] V.B. Berestetskii, E.M. Lifshitz and L.P. Pitaevskii, *Relativistic Quantum Theory*, Vol. **4**, Part I of *Course of Theoretical Physics*, Section 34, Pergamon Press, Oxford (1971).

[5] A.A. Sokolov, I.M. Ternov and V.Ch. Zhukovskii, *Quantum Mechanics*, Mir Publishers, Moscow (1984).

12 Identical Particles

12.1 Introduction

In classical mechanics, identical particles are distinguished by their position and momentum, i.e. by their path. However, on account of uncertainty principle this is not possible in quantum mechanics and hence identical particles lose their distinguishability. This gives rise to a new conservation law under interchange or permutation of any two particles k and l, defined by

$$P_{kl}\psi(1, 2, \ldots, k, \ldots, l \ldots) = \psi(1, 2, \ldots, l \ldots k \ldots). \tag{12.1}$$

This is expressed by invariance of the Hamiltonian under this transformation:

$$[P_{kl}H] = 0. \tag{12.2}$$

For a system of two identical particles the eigenvalue equation for the permutation operator can be written as

$$P_{12}\psi(1, 2) = \lambda\psi(1, 2) \tag{12.3}$$

from which we get, by repeated operations,

$$P_{12}^2\psi(1, 2) = \lambda^2\psi(1, 2). \tag{12.4}$$

Also, by definition (12.1),

$$P_{12}^2\psi(1, 2) = P_{12}\psi(21) = \psi(1, 2). \tag{12.5}$$

From (12.4) and (12.5) we have

$$\lambda^2 = 1, \quad \lambda = \pm 1. \tag{12.6}$$

The eigenfunctions ψ_s and ψ_a corresponding to eigenvalues $(+1)$ and (-1) are called symmetric and antisymmetric respectively. These two states can be represented by

$$\psi_s = \frac{1}{\sqrt{2}}[\psi(1, 2) + \psi(2, 1)],$$

$$\psi_a = \frac{1}{\sqrt{2}}[\psi(1, 2) - \psi(2, 1)]. \tag{12.7}$$

For N particles,

$$\psi_s = N_s \sum_n P_n \psi(1, 2, \dots N),$$

$$\psi_a = N_a \sum_n (-1)^n P_n \psi(1, 2, \dots N), \tag{12.8}$$

where $P_n\psi(1, 2, \dots , N)$ is a function obtained from $\psi(1, 2, \dots , N)$ by consecutive permutation of pairs of particles. This also holds for a many-particle system when any two particles are interchanged as stated earlier. Particles constituting systems whose permutation eigenfunction is symmetric are called 'bosons' and those which are antisymmetric are called 'fermions' named after Satyendranath Bose and Enrico Fermi respectively, who gave the laws of energy distribution for these particles. There exists a connection between spin and statistics which has been proved in the framework of relativistic quantum field theory according to which particles with half-odd integral spin are fermions while those with integral spin are bosons. It is also true for composites like atoms and atomic nuclei. Atoms whose total number of electrons and protons and neutrons is odd obey Fermi statistics and those for which this is even obey Bose statistics. For atomic nuclei, the same holds for the sum of number of protons and neutrons.

12.2 Pauli Exclusion and Bose Aggregation Principle

For a system of N non-interacting particles, the Hamiltonian will be a sum of those of single particles $H(i)$

$$H = \sum_i^N H(i) \tag{12.9}$$

and the wavefunction $\psi(1, 2, \dots N)$ will, therefore, be a product of single particle wavefunctions $\phi_{n_1}(1)\ \phi_{n_2}(2) \dots \phi_{nN}(N)$ obeying the Schrödinger equation

$$H(i)\phi_{n_i}(i) = \epsilon_{n_i}\phi_{n_i}(i), \tag{12.10}$$

where n_i are the quantum numbers of single-particle states. The symmetrical and antisymmetrical wavefunctions given in (12.8) can, then, be written as

$$\psi_s = N \sum_n P_n \phi_{n_1}(1)\phi_{n_2}(2) \dots \phi_{n_N}(N), \tag{12.11}$$

$$\psi_a = \frac{1}{\sqrt{N!}} \sum_n (-1)^n P_n \phi_{n_1}(1)\phi_{n_2}(2) \dots \phi_{n_N}(N). \tag{12.12}$$

The latter can be expressed in the form of a determinant, the Slater determinant,

$$\psi_a = \frac{1}{\sqrt{N!}} \begin{vmatrix} \phi_{n_1}(1) & \phi_{n_1}(2) & \dots & \phi_{n_1}(N) \\ \phi_{n_2}(1) & \phi_{n_2}(2) & \dots & \phi_{n_2}(N) \\ \dots & \dots & \dots & \dots \\ \dots & \dots & \dots & \dots \\ \phi_{n_N}(1) & \phi_{n_N}(2) & \dots & \phi_{n_N}(N) \end{vmatrix}. \tag{12.13}$$

If any two particles are in the same state, for instance

$$\phi_{n_1}(1) = \phi_{n_2}(2), \tag{12.14}$$

then the first two rows of the determinant will be same and the determinant will vanish. It will be non-zero only when all the quantum numbers, $n_1, n_2 \dots n_N$, are different. This is the Pauli exclusion principle which states that in a system of identical fermions, two or more particles cannot be in the same state at the same time. It is as if fermions repel or avoid each other. For a two-particle system, for example,

$$\psi_a = \frac{1}{\sqrt{2}} \begin{vmatrix} \phi_{n_1} & \phi_{n_1}(2) \\ \phi_{n_2} & \phi_{n_2}(2) \end{vmatrix} = \frac{1}{\sqrt{2}} [\phi_{n_1}(1)\phi_{n_2}(2) - \phi_{n_2}(1)\phi_{n_1}(2)]. \tag{12.15}$$

If the two particles (1) and (2) are in the same state, i.e., $n_2 = n_1$, $\psi_a = 0$. Thus the probability for the two-particles to be in the same state is zero.

For bosons on the other hand, there are three symmetrical states for the two-particle system:

$$\begin{aligned} \psi_s^{\mathrm{I}} & \frac{1}{\sqrt{2}} [\phi_{n_1}(1)\phi_{n_2}(2) + \phi_{n_2}(1)\phi_{n_1}(2)] \\ \psi_s^{\mathrm{II}} &= \phi_{n_1}(1)\phi_{n_1}(2) \\ \psi_s^{\mathrm{III}} &= \phi_{n_2}(1)\phi_{n_2}(2) \end{aligned} \tag{12.16}$$

Since all three states are equally probable, it follows that the probability for the two particle (1) and (2) to be in the same state is $\frac{1}{3}$, which is greater than that of classical theory of $\frac{1}{4}$. Thus, the tendency of indistinguishable particles to be together is greater in symmetrical quantum state than in classical state. They avoid to be together in antisymmetrical state. The former may be called Bose aggregation principle just as the latter is called Pauli exclusion principle.

Another way of looking at this problem is through matrix elements of creation and annihilation operators introduced later in this chapter in sections 4 and 5 where it is shown that for bosons,

$$\langle n+1|a^{\dagger}|n\rangle = \sqrt{n+1},$$

which shows that the probability of creation of an additional boson in a state with n bosons is $(n+1)$ whereas for fermions, where $n = 1, 0$, it is either 1 or zero.

12.3 Exchange Interaction

A peculiar dependence of the energy of a system of particles on its total spin occurs due to indistinguishability which goes by the name of exchange interaction. To see how this comes about, we consider two free electrons which have spin $\frac{1}{2}$ and are fermions on account of which their total wavefunction

$$\Psi(r_1 s_1, r_2 s_2) = \phi(\vec{r}_1 \vec{r}_2)\chi(1,2), \tag{12.17}$$

which is a product of space part ϕ and a spin part χ, has to be antisymmetric. The total spin S of the two electrons can be $S = 0$, and $S = 1$. The corresponding eigenfunctions are

$$\chi_{S=0,m_S=0} = \frac{1}{\sqrt{2}}[\alpha(1)\beta(2) - \alpha(2)\beta(1)], \tag{12.18}$$

$$\chi_{S=1,m_S=0} = \frac{1}{\sqrt{2}}[\alpha(1)\beta(2) + \alpha(2)\beta(1)],$$

$$\chi_{S=1,m_S=1} = \alpha(1)\alpha(2), \tag{12.19}$$

$$\chi_{S=1,m_S=-1} = \beta(1)\beta(2),$$

where α and β are wavefunctions of a single electron with spin up and down respectively. It will be noticed that while the spin wavefunction (12.18) for $S = 0$ is antisymmetric with respect to interchange of the electrons, the three wavefunctions (12.19) corresponding to $S = 1$ are symmetric. Since the total wavefunction Ψ has to be antisymmetric, $\phi(\vec{r}_1 \vec{r}_2)$ for $S = 0$ has to be symmetric and for $S = 1$, it has to be antisymmetric. In the absence of interaction between the two electrons there is a degeneracy—the energy eigenvalues of these two states are equal. This is called exchange degeneracy.

When there is interaction between the two electrons, represented by the potential $V(\vec{r}_1 \vec{r}_2)$, the first order correction to energy is given by its expectation value for

$$\Phi_{s,a} = [\phi_1(r_1)\phi_2(r_2) \pm \phi_1(r_2)\phi_2(r_1)]. \tag{12.20}$$

For the symmetric state corresponding to $S = 0$,

$$\langle\phi_S|V|\phi_s\rangle = J + K, \tag{12.21}$$

where

$$J = \int d^3r_1 \int d^3r_2 |\phi_1(r_1)|^2 V(r_1r_2)|\phi_2(r_2)|^2$$

and

$$K = \int d^3r_1 \int d^3r_2 \phi_1(r_1)\phi_1^*(r_2)V(r_1r_2)\phi_2(r_2)|\phi_2^*(r_1), \tag{12.22}$$

while for the antisymmetric state corresponding to $S = 1$, it is

$$\langle\phi_a|V|\phi_a\rangle = J - K. \tag{12.23}$$

Thus there arises an energy difference

$$\Delta E = 2K \tag{12.24}$$

between $S = 0$ and $S = 1$ states. The integral K is called the 'exchange integral'. If the interaction V is Coulombic as in the He atom, the integral J is called the Coulomb integral.

12.4 Fock Space (Second) Quantization

An elegant method to deal with a system of indistinguishable particles was presented by Fock by introducing the concept of occupation number space. State vectors $\Phi(k_1k_2\ldots k_n)$ specified by momenta of individual particles are constructed by application of creation operators $a^\dagger(k)$ operating on the vacuum state $|0>$:

$$\Phi(\vec{k}_1\vec{k}_2\ldots\vec{k}_n) = \frac{1}{\sqrt{n!}}\, a^\dagger(k_n)\ldots a^\dagger k_1|0> \tag{12.25}$$

where, the vacuum state is defined by

$$a(k)|0> = a_k|0> = 0. \tag{12.26}$$

The $a_k^\dagger$ and a_k obey commutation rules

$$[a_k, a_{k'}] = [a_k^\dagger, a_{k'}^\dagger] = 0,\ [a_k, a_{k'}^\dagger] = \delta_{kk'} \tag{12.27}$$

In terms of these operators, the energy, momentum and number operators are,

$$H = \sum_k \epsilon_k a_k^\dagger a_k, \quad \vec{P} = \sum_k \vec{k} a_k^\dagger a_k,$$

$$N = \sum_k a_k^\dagger a_k = \sum_k n_k, \tag{12.28}$$

$\epsilon_{k_i}, \vec{k}_i$ being single-particle energy and momenta.

It is easy to see by using commutation relation (12.27) that

$$\begin{aligned}
Ha_k^\dagger\Phi(\vec{k}_1\vec{k}_2\ldots\vec{k}_n) &= (\epsilon_k+\epsilon_{k_1}+\epsilon_{k_2}+\cdots+\epsilon_{k_n})a_k^\dagger\Phi(\vec{k}_1\vec{k}_2\ldots\vec{k}_n)\\
\vec{P}a_k^\dagger\Phi(\vec{k}_1\vec{k}_2\ldots\vec{k}_n) &= (\vec{k}+\vec{k}_1+\vec{k}_2+\cdots+\vec{k}_n)a_k^\dagger\Phi(\vec{k}_1\vec{k}_2\ldots\vec{k}_n)\\
Na_k^\dagger\Phi(\vec{k}_1\vec{k}_2\ldots\vec{k}_n) &= (n+1)a_k^\dagger\Phi(\vec{k}_1\vec{k}_2\ldots\vec{k}_n) \qquad (12.29)
\end{aligned}$$

and

$$\begin{aligned}
Ha_k\Phi(\vec{k}_1\vec{k}_2\ldots\vec{k}_n) &= (-\epsilon_k+\epsilon_{k_1}+\epsilon_{k_2}+\cdots+\epsilon_{k_n})a_k\Phi(\vec{k}_1\vec{k}_2\ldots\vec{k}_n)\\
\vec{P}a_k\Phi(\vec{k}_1\vec{k}_2\ldots\vec{k}_n) &= (-\vec{k}+\vec{k}_1+\vec{k}_2+\cdots+\vec{k}_n)a_k\Phi(\vec{k}_1,\vec{k}_2\ldots,\vec{k}_n)\\
Na_k\Phi(\vec{k}_1\vec{k}_2\ldots\vec{k}_n) &= (n-1)a_k\Phi(\vec{k}_1\vec{k}_2\ldots\vec{k}_n), \qquad (12.30)
\end{aligned}$$

where

$$\begin{aligned}
H\Phi(\vec{k}_1,\vec{k}_2,\ldots,\vec{k}_n) &= (\epsilon_{k_1}+\epsilon_{k_2}+\cdots+\epsilon_{k_n})\Phi(\vec{k}_1,\vec{k}_2,\ldots,\vec{k}_n)\\
P\Phi(\vec{k}_1,\vec{k}_2,\ldots,\vec{k}_n) &= (\vec{k}_1+\vec{k}_2+\cdots+\vec{k}_n)\Phi(\vec{k}_1,\vec{k}_2,\ldots,\vec{k}_n)\\
N\Phi(\vec{k}_1,\vec{k}_2,\ldots,\vec{k}_n) &= (n_{k_1}+n_{k_2}+\cdots+n_{k_n})\Phi(\vec{k}_1,\vec{k}_2,\ldots,\vec{k}_n)\\
(n_{k_1}+n_{k_2}+\cdots+n_{k_n}) &= n.
\end{aligned}$$

It is to be noted that $\vec{k}$ in (12.30) should equal any one of $\vec{k}_1\vec{k}_2\ldots\vec{k}_n$, otherwise $a_k\Phi(\vec{k}_1\ldots\vec{k}_n)$ will be zero. Equation (12.29) leads to

$$a_k^\dagger\Phi(\vec{k}_1\ldots\vec{k}_n) = \Phi(\vec{k},\vec{k}_1\ldots\vec{k}_n), \qquad (12.31)$$

which tells us that $a_k^\dagger$ acting $\Phi(\vec{k}_1\ldots\vec{k}_n)$ has transformed it to a state with one more particle with energy and momentum $\epsilon_{\vec{k}}$ and $\vec{k}$. Similarly, (12.30) tells us that a_k transforms the same state to a state with one less particle. Therefore these are called creation and annihilation operators respectively.

The connection of the above Fock space formulation, which is in the Heisenberg picture, with the Schrödinger picture can be obtained by going over to the configuration space field operator $\phi(x)$ in which (12.25) reads

$$\Phi(\vec{r}_1\vec{r}_2\ldots\vec{r}_n) = \frac{1}{\sqrt{n!}}\phi^\dagger(\vec{r}_1)\ldots\phi^\dagger(\vec{r}_n)|0> \qquad (12.32)$$

with

$$\phi(\vec{r}) = \frac{1}{(2\pi)^{3/2}}\int d^3k e^{i\vec{k}\ldots\vec{r}}a_k \qquad (12.33)$$

which can be called the Schrödinger field operator. The Schrödinger state vector defined by

$$\Phi(t) = \sum_{n=0}^{\infty} \int d^3 r_1 \ldots d^3 r_n \psi(\vec{r}_1, \vec{r}_2 \ldots \vec{r}_n t) \Phi(\vec{r}_1, \vec{r}_2 \ldots \vec{r}_n) \tag{12.34}$$

obeys the equation

$$H\Phi(t) = i\frac{\partial \Phi}{\partial t} \tag{12.35}$$

if

$$\left(i\frac{\partial}{\partial t} + \frac{1}{2m}\sum_{i=1}^{n} \nabla_i^2 \right) \psi(\vec{r}_1 \vec{r}_2 \ldots \vec{r}_n t) = 0, \tag{12.36}$$

which is the many-particle Schrödinger equation. The energy, momentum and number operators defined in (12.28) written in terms of $\phi(xt)$ have the forms

$$\begin{aligned} H &= \int d^3 r \partial_i \phi^+ \partial_i \phi(\vec{r}), \\ P_i &= \frac{i}{2} \int d^3 r [\partial_i \phi^+(\vec{r}t)\phi(\vec{r}t) - \phi^+(\vec{r}t)\partial_i \phi(\vec{r}t)], \\ N &= \int d^3 r \phi^+(\vec{r}t)\phi(\vec{r}t), \end{aligned} \tag{12.37}$$

and the commutation relations in (12.27) become

$$\begin{aligned} [\phi(\vec{r}t), \phi^+(\vec{r}'t)] &= \delta(\vec{r} - \vec{r}'), \\ [\phi(\vec{r}t), \phi(\vec{r}'t)] &= [\phi^+(\vec{r}, t)\phi^+(\vec{r}', t)] = 0. \end{aligned} \tag{12.38}$$

The indistinguishable many-particle system can thus be described in terms of operators (12.37) and the commutation relations (12.38) which are written in terms of Schrödinger field operators $\phi^+(\vec{r}t)$ and $\phi(\vec{r}t)$. This is the second quantized Schrödinger field theory. The first quantized Schrödinger theory describes dynamics of single particles.

The commutation rule

$$\phi^+(\vec{r}_1)\phi^+(\vec{r}_2) - \phi^+(\vec{r}_2)\phi^+(\vec{r}_1) = 0,$$

is equivalent to

$$P\psi(1, 2) = \psi(2, 1) \tag{12.39}$$

for a symmetric state of two particles. Thus the above treatment is valid for bosons. For fermions, H, $\vec{P}$ and N remain the same, but the commutation relations are replaced by anti commutation rules

$$\{a_k, a_{k'}\} = \{a_k^\dagger, a_{k'}^\dagger\} = 0, \;\; \{a_k, a_{k'}^+\} = \delta_{kk'}, \tag{12.40}$$

which in configuration space read as

$$\{\phi(\vec{r}), \phi(\vec{r}')\} = \{\phi^\dagger(\vec{r}), \phi^\dagger(\vec{r}')\} = 0, \;\; \{\phi(\vec{r}), \phi^\dagger(\vec{r}')\} = \delta(\vec{r} - \vec{r}'). \tag{12.41}$$

The first two of the above are equivalent to

$$P\psi(1,2) = -\psi(2,1) \tag{12.42}$$

for the antisymmetric state of two particles. Thus Fock space quantization with anti commutation rule hold for indistinguishable fermions. Further, if we put $\vec{r}_1 = \vec{r}_2 = \vec{r}$ in the first two equation of (12.41), we get $(\phi(\vec{r}t))^2 = (\phi^\dagger(\vec{r}t))^2 = 0$, which means that two spinless fermions cannot occupy the same position.

12.5 Quantization of the Electromagnetic Field

The electromagnetic field in free space represented by Maxwell equations (with $c = 1$)

$$\vec{\nabla} \cdot \vec{E} = 0, \quad \vec{\nabla} \times \vec{B} - \frac{\partial \vec{E}}{\partial t} = 0$$

$$\vec{\nabla} \cdot \vec{B} = 0, \quad \vec{\nabla} \times \vec{E} + \frac{\partial \vec{B}}{\partial t} = 0 \tag{12.43}$$

can be written in terms of the vector potential $\vec{A}$ defined by

$$\vec{E} = -\frac{\partial \vec{A}}{\partial t} \qquad \vec{B} = \vec{\nabla} \times \vec{A} \tag{12.44}$$

as

$$\nabla^2 \vec{A} - \frac{\partial^2 \vec{A}}{\partial t} = 0, \quad \vec{\nabla} \cdot \vec{A} = 0. \tag{12.45}$$

The potential $\vec{A}$ can be expanded in plane waves:

$$\vec{A}(\vec{r}) = \sum_{k,\lambda} \frac{1}{\sqrt{2\omega_k}} \vec{e}_{(\lambda)} (a_{k,\lambda} e^{i\vec{k}\cdot\vec{r} - i\omega_k t} + c.c), \tag{12.46}$$

where $\vec{e}^{(\lambda)}$ $(\lambda = 1, 2)$ are the polarisation unit vectors and $a_{k\lambda}$ are the amplitudes of plane waves. From (12.45) and (12.46), it follows that

$$\vec{E} = \sum_{k,\lambda} i\vec{e}_{(\lambda)}\sqrt{\frac{\omega_k}{2}}(a_{k,\lambda}e^{i\vec{k}\cdot\vec{r}-i\omega_k t} - c.c),$$

$$\vec{B} = \sum_{k,\lambda} \frac{i\vec{k}\times\vec{e}_{(\lambda)}}{\sqrt{2\omega_k}}(a_{k,\lambda}e^{i\vec{k}\cdot\vec{r}-i\omega_k t} - c.c). \tag{12.47}$$

The energy of the field then works out to

$$\epsilon = \frac{1}{2}\int dr^3(E^2 + B^2) = \sum_{k,\lambda}\omega_k(a_{k\lambda}a^*_{k\lambda}). \tag{12.48}$$

In terms of

$$p_{k\lambda} = \sqrt{\frac{\omega_k}{2}}(a_{k\lambda} + a^*_{k\lambda}),$$

$$q_{k\lambda} = \frac{i}{\sqrt{2\omega_k}}(a^*_{k\lambda} - a_{k\lambda}), \tag{12.49}$$

the energy of the field has the form

$$\epsilon = \sum_{k\lambda}\left(\frac{p^2_{k\lambda}}{2} + \frac{1}{2}\omega_k^2 q^2_{k\lambda}\right), \tag{12.50}$$

which is the energy of an assembly of harmonic oscillators. The electromagnetic field therefore can be considered as such.

The second quantization of the field can now be done in the manner in which it is done for harmonic oscillator, i.e.,

$$[a_{k\lambda}, a^\dagger_{k'\lambda'}] = \delta_{kk'}\delta_{\lambda\lambda'},$$

$$[a_{k\lambda}, a_{k'\lambda'}] = [a^\dagger_{k\lambda}, a^\dagger_{k'\lambda'}] = 0, \tag{12.51}$$

where $a_{k\lambda}$ are now operators with properties (see (7.78))

$$a^\dagger_{k\lambda}|n_{k\lambda}\rangle = (n_{k\lambda}+1)^{1/2}|n_{k\lambda}+1\rangle$$

$$a_{k\lambda}|n_{k\lambda}\rangle = n^{1/2}_{k\lambda}|n_{k\lambda}-1\rangle \tag{12.52}$$

$$N|n_{k\lambda}\rangle = a^\dagger_{k\lambda}a_{k\lambda}|n_{k\lambda}\rangle = n_{k\lambda}|n_{k\lambda}\rangle.$$

The quantization (12.51) makes $\vec{A}(\vec{r})$ a second quantized operator with matrix elements

$$\langle n_{k\lambda}-1|\vec{A}(\vec{r})|n_{k\lambda}\rangle = n^{1/2}_{k\lambda}\frac{\vec{e}_{(\lambda)}}{\sqrt{2\omega_k}}e^{i\vec{k}\cdot\vec{r}-i\omega_k t}$$

$$\langle n_{k\lambda}+1|\vec{A}(\vec{r})|n_{k\lambda}\rangle = (n_{k\lambda}+1)^{1/2}\frac{\vec{e}_{(\lambda)}}{\sqrt{2\omega_k}}e^{-i\vec{k}\cdot\vec{r}+i\omega_k t}. \tag{12.53}$$

12.6 Parabosons and Parafermions

We have seen in the previous sections that Schrödinger fields obeying commutation relation represent bosons while those obeying anticommutation relation represent fermions. This result can be derived from a combination of Schrödinger equation

$$i\frac{\partial\phi}{\partial t} = -\frac{1}{2m}\nabla^2\phi \tag{12.54}$$

and the Heisenberg equation

$$i\frac{\partial\phi}{\partial t} = [\phi, H], \tag{12.55}$$

which gives

$$[\phi, H] = -\frac{1}{2m}\nabla^2\phi, \tag{12.56}$$

and which in momentum space reads $[a_k, H] = (k^2/2m)a_k = E_k a_k$. If we use the symmetrised Hamiltonian

$$\begin{aligned} H &= \frac{1}{4m}\int d^3x(\vec{\nabla}\phi^+ \cdot \vec{\nabla}\phi + \vec{\nabla}\phi \cdot \vec{\nabla}\phi)^+ \\ &= \frac{1}{2}\sum_l E_l(a_l^+ a_l + a_l a_l^+) \end{aligned} \tag{12.57}$$

in (12.56), we get

$$\begin{aligned} &[a_{k'}, (a_l^+ a_l + a_l a_l^+)] = 2\delta_{kl} a_l \\ &[a_{k'}^+ (a_l^+ a_l + a_l a_l^+)] = -2\delta_{kl} a_l^+. \end{aligned} \tag{12.58}$$

This will be satisfied if

$$[a_k, a_l^+] = \delta_{kl}, \quad [a_k, a_l] = [a_k^+, a_l^+] = 0. \tag{12.59}$$

However, in general, (12.58) will be satisfied if

$$\begin{aligned} &[a_k, \{a_l^+, a_m\}] = 2\delta_{kl} a_m \\ &[a_k, \{a_l^+, a_m^+\}] = 2\delta_{kl} a_m^+ + 2\delta_{km} a_l^+ \\ &[a_k, \{a_l, a_m\}] = 0. \end{aligned} \tag{12.60}$$

These are known as parabose commutation rules and the particles which the fields ϕ represent are known as parabosons. In configuration space these commutation rules read as

$$
\begin{aligned}
[\phi(\vec{x}), \{\phi^+(\vec{y}), \phi(\vec{z})\}] &= 2\delta(\vec{x}-\vec{y})\phi(\vec{z}) \\
[\phi(\vec{x}), \{\phi^+(\vec{y}), \phi^+(\vec{z})\}] &= 2\delta(\vec{x}-\vec{y})\phi(\vec{z}) + 2\delta(\vec{x}-\vec{z})\phi(\vec{y}) \\
[\phi(\vec{x}), \{\phi(\vec{y}), \phi(\vec{z})\}] &= 0.
\end{aligned} \tag{12.61}
$$

On the other hand, if we use the antisymmetrised Hamiltonian

$$
\begin{aligned}
H &= \frac{1}{4m}\int d^3x(\vec{\nabla}\phi^+ \cdot \vec{\nabla}\phi - \vec{\nabla}\phi \cdot \vec{\nabla}\phi^+) \\
&= \frac{1}{2}\sum_l E_l(a_l^+ a_l - a_l a_l^+)
\end{aligned} \tag{12.62}
$$

in (2.56), we would get

$$
\begin{aligned}
[a_k, (a_l^+ a_l - a_l a_l^+)] &= 2\delta_{kl} a_l \\
[a_k^+, (a_l^+ a_l - a_l a_l^+)] &= -2\delta_{kl} a_l^+.
\end{aligned} \tag{12.63}
$$

This will be satisfied if

$$
\{a_k a_l^+\} = \delta_{kl} \qquad \{a_k a_l\} = \{a_k^+ a_l^+\} = 0. \tag{12.64}
$$

As in the symmetrised Hamiltonian case, the general commutation rule will be

$$
\begin{aligned}
[a_k, [a_l^+, a_m]] &= 2\delta_{kl} a_m \\
[a_k, [a_l^+, a_m^+]] &= 2\delta_{kl} a_m^+ - 2\delta_{km} a_l^+ \\
[a_k, [a_l, a_m]] &= 0.
\end{aligned} \tag{12.65}
$$

These are known as parafermi commutation rules and particles that obey these are called parafermions. In configuration space these become

$$
\begin{aligned}
[\phi(\vec{x}), [\phi^+(\vec{y}), \phi(\vec{z})]] &= 2\delta(\vec{x}-\vec{y})\phi(\vec{z}) \\
[\phi(\vec{x}), [\phi^+(\vec{y}), \phi^+(\vec{z})]] &= 2\delta(\vec{x}-\vec{y})\phi(\vec{z}) - 2\delta(\vec{x}-\vec{z})\phi(\vec{y}) \\
[\phi(\vec{x})\, [\phi(\vec{y}), \phi(\vec{z})]] &= 0.
\end{aligned} \tag{12.66}
$$

It is to be noted that all the fields appearing in the configuration space commutators are to be taken at equal times.

Equations (12.60) and (12.65) are general commutation rules which are obtained from the requirement that the fields satisfy both Schrödinger and Heisenberg equations of motion. For the specific case of two-particle states with momenta $\vec{k}$ and $\vec{l}$, it is possible to get

$$
\begin{aligned}
[a_k, a_l^+]_\pm &= \delta_{kl} \\
[a_k, a_l]_\pm &= [a_k^+, a_l^+]_\pm = 0.
\end{aligned} \tag{12.67}
$$

The plus sign comes from (12.65) and minus sign from (12.60). These are the Fermi and Bose commutation rules. From the second and third commutation rules with plus sign, it can be seen that if $\vec{k} = \vec{l}$, $(a_{\bar{k}})^2 = 0$, $(a_k^+)^2 = 0$, which tells us that one cannot put two fermions in one state; only one particle can occupy one state. This corresponds to occupation number $p = 1$.

For three-particle states with momenta $\vec{k}, \vec{l}, \vec{m}$, one obtains (we shall not get into the derivation here)

$$\begin{aligned} a_k a_l^+ a_m \pm a_m a_l^+ a_k &= 2\delta_{kl} a_m \pm 2\delta_{klm} a_k \\ a_k a_l a_m^+ \pm a_m^+ a_l a_m &= 2\delta_{lm} a_k \\ a_k a_l a_m \pm a_m a_l a_k &= 0, \end{aligned} \tag{12.68}$$

the plus sign standing for parafermions and the minus for parabosons. It will be seen from the third commutator that for the parafermi case, if $\vec{k} = \vec{l} = \vec{m}$, then $(a_k)^3 = 0 = (a_k^+)^3$, which tells us that one cannot put three parafermion in a given state. One can put at best two such particles in one state. This corresponds to $p = 2$. For parafermions obeying commutation relation of arbitrary order p, there can be maximum of p particles in a given state. For parabosons there is no such restriction; one can put any number of particles in a given state.

Problems

1. As considered in the text, space wave function of two spin-$\frac{1}{2}$ electrons can be either symmetric or antisymmetric under the exchange of their space coordinates. The energy eigenvalues corresponding to these two wave functions are equal in the absence of interaction between them, but when they interact there is an energy difference. Suppose at $t = 0$ their wavefunctions is

$$\psi(\vec{r}_1, \vec{r}_2) = \phi_1(\vec{r}_1)\phi_2(\vec{r}_2).$$

Obtain expression for $\psi(\vec{r}_1, \vec{r}_2, t)$ at a later time and show that the two electrons exchange their positions periodically. Calculate the time period as a function of the energy difference between the symmetric and anti-symmetric states. This is known as exchange oscillation, quite similar to Stark oscillations discussed in chapter 10.

Bibliography

[1] V. Fock, *Z. Phys.*, **49**, 339 (1928); *Phys. Phy. Z. Sowjet USSR*, **6**, 425 (1934).

[2] P.A.M. Dirac, *Proc. Roy. Soc.*, A**112**, 661 (1926).

[3] S.N. Bose, *Z. Physik*, **26**, 178 (1924).
[4] E. Fermi, *Z. Physik*, **36**, 902 (1926).
[5] G. Gentile, *Nuovo Cimento*, **17**, 493 (1940); **19**, 106 (1942).
[6] Y. Ohnuki and S. Kamefuchi, *Quantum Field Theory and Parastatistics*, Springer Verlag, New York (1982), Chapter 5.
[7] Y. Takahashi, *An Introduction to Field Quantization*, Pergamon Press, Oxford (1969).

13 The Helium Atom

13.1 Ground State

The helium atom has a nucleus of two protons and two neutrons with charge $+2e$ and two atomic electrons. If we ignore the electron spin, the Hamiltonian can be written as

$$H = \frac{-\hbar^2}{2m}(\nabla_1^2 + \nabla_2^2) - 2e^2\left(\frac{1}{r_1} + \frac{1}{r_2}\right) + \frac{e^2}{|\vec{r}_1 - \vec{r}_2|}, \tag{13.1}$$

where $\vec{r}_1$ and $\vec{r}_2$ are the coordinates of the two electrons. In the ground state, the two electrons are in s-orbitals $(2s^2)$. If we neglect the third term in (13.1) which represents the repulsive coulomb interaction between the two electrons, the wavefunction can be written as product of two hydrogenic wave functions

$$\psi_0 = \phi_{1s}(1)\phi_{1s}(2) = \frac{8}{\pi a_0^3} e^{-\frac{2}{a_0}(r_1+r_2)}, \tag{13.2}$$

where $a_0 = h^2/me^2$. The energy of the ground state in this approximation then works out to

$$E_0 = -\frac{4e^2}{a_0}. \tag{13.3}$$

The wavefunction (13.2) is symmetric in r_1 and r_2. The antisymmetric one is identically zero. The spin wavefunction therefore has to be antisymmetric, i.e. the two spins are antiparallel (singlet) and spectroscopic notation for the state is $^1S_0$. Such a state with antiparallel spins is called a 'para' state.

The contribution of the electron–electron interaction term can be calculated using perturbation theory. In the first order of approximation,

$$\Delta E = e^2 \int d^3r_1, \int d^3r_2 \psi_0^*(\vec{r}_1, \vec{r}_2) \frac{1}{|\vec{r}_1 - \vec{r}_2|} \psi_0(\vec{r}_1, \vec{r}_2). \tag{13.4}$$

Expanding the interaction potential energy term in spherical harmonics,

$$\frac{1}{|\vec{r}_1 - \vec{r}_2|} = \begin{cases} \frac{4\pi}{r_1} \sum_{l,m} \frac{1}{2l+1} \left(\frac{r_2}{r_1}\right)^l Y^*_{lm}(\theta_1\varphi_1) Y_{lm}(\theta_2\varphi_2) & \text{if } r_1 > r_2, \\ \frac{4\pi}{r_2} \sum_{l,m} \frac{1}{2l+1} \left(\frac{r_1}{r_2}\right)^l Y^*_{lm}(\theta_1\varphi_1) Y_{lm}(\theta_2\varphi_2) & \text{if } r_2 > r_1, \end{cases} \tag{13.5}$$

and substituting (13.5) in (13.4) yields

$$\begin{aligned} \Delta E &= 16\left(\frac{2}{a_0}\right)^6 \int_0^\infty dr_1 r_1^2 e^{-4r_1/a_0} \\ &\quad \times \left[\frac{1}{r_1}\int_0^{r_1} dr_2 r_2^2 e^{-4r_2/a_0} + \int_{r_1}^\infty dr_2 r_2 e^{-4r_2/a_0}\right] \qquad (13.6) \\ &= \frac{5e^2}{4a_0}. \end{aligned}$$

Therefore,

$$E = E_0 + \Delta E = \left(-4 + \frac{5}{4}\right)\frac{e^2}{a_0} = -\frac{11e^2}{4a_0} = -2.75\frac{e^2}{a_0}. \tag{13.7}$$

This is to be compared with the experimental value $E_{\text{expt}} = -2.904\frac{e^2}{a_0}$.

Instead of using perturbation theory, if we make a variational calculation using trial wavefunction with β as a parameter,

$$\psi_0 = \frac{1}{\pi}\left(\frac{\beta}{a_0}\right)^3 e^{-\beta(r_1+r_2)/a_0}, \tag{13.8}$$

we get

$$E(\beta) - \frac{e^2}{a_0}\left(\beta^2 - \frac{27}{8}\beta\right). \tag{13.9}$$

Minimising E as a function of the variational parameter β, i.e.

$$\frac{dE(\beta)}{d\beta} = 0, \tag{13.10}$$

gives

$$\beta = \frac{27}{16}. \tag{13.11}$$

Substituting this in (13.9) and (13.8) gives

$$E = \frac{-2.85e^2}{a_0}, \tag{13.12}$$

$$\psi_0 = \frac{1}{\pi}\left(\frac{27}{16a_0}\right)^3 e^{-\frac{27}{16a_0}(r_1+r_2)}. \tag{13.13}$$

As already noted, $E_{\text{expt}} = -2.904e^2/a_0$.

For helium-like ions such as singly-ionized lithium atom and doubly-ionized berilium atom with nuclear charge Z, we get from perturbation theory,

$$E = \frac{-Ze^2}{a_0}\left(Z - \frac{5}{8}\right)$$

and from variational calculation,

$$E = \left[Z^2 - \frac{5}{8}Z + \frac{25}{256}\right], \tag{13.14}$$

$$\psi_0 = \frac{1}{\pi}\left(\frac{Z^*}{a_0}\right)^3 e^{-Z^*(r_1+r_2)/a_0}, \tag{13.15}$$

where $Z^* = Z - \frac{5}{16}$, which can be regarded as effective charge and can be thought of as due to screening of the nuclear charge for one electron by the other.

13.2 First Excited State

The first excited state has one electron in the $1s$ orbital whereas the other is in the $2s$. Therefore, one can have space-symmetric and antisymmetric wavefunctions

$$\psi_s = \frac{1}{\sqrt{2}}\left[\phi_{1s}(1)\phi_{2s}(2) + \phi_{1s}(2)\phi_{2s}(1)\right], \tag{13.16}$$

$$\psi_a = \frac{1}{\sqrt{2}}\left[\phi_{1s}(1)\phi_{2s}(2) - \phi_{1s}(2)\phi_{2s}(1)\right], \tag{13.17}$$

$$E_2 = \epsilon_{1s} + \epsilon_{2s},$$

both having unperturbed energy eigenvalue.

Since the product of space and spin wavefunctions has to be antisymmetric, ψ_s will have antiparallel spin whereas ψ_a will have parallel spin. The former is called 'para' state with spectroscopic notation ${}^1S_0$ and the latter as 'ortho' state with notation ${}^3S_1$. Because of selection rule which prohibits radiative transition from triplet to singlet states, helium atoms, once they are in a singlet state, can never go to triplet state by radiative transition and vice versa. For this reason, it was thought for some time that there are two forms of helium—ortho and para helium. It is worth mentioning here that such transitions are allowed through spin–orbit interaction and have been observed in heavier atoms with high value of nuclear charge.

We have already seen in the previous chapter that when there is exchange degeneracy resulting in two wavefunctions for a given eigenvalue as in (13.16) and (13.17), the Coulomb interaction between the two electrons will lead to the lifting of the degeneracy giving energy for para and ortho states as

$$E_s = \epsilon_{1s} + \epsilon_{2s} + J + K, \tag{13.18}$$

$$E_a = \epsilon_{1s} + \epsilon_{2s} + J - K, \tag{13.19}$$

where

$$J = \int d^3r_1 \int d^3r_2 |\phi_{1s}(r_1)|^2 \frac{Ze^2}{|\vec{r}_1 - \vec{r}_2|} |\phi_{2s}(r_2)|^2 \tag{13.20}$$

is called the Coulomb integral and

$$K = \int d^3r_1 \int d_2^3 \phi_{1s}(r_1)\phi_{1s}^*(r_2) \frac{Ze^2}{|\vec{r}_1 - \vec{r}_2|} |\phi_{2s}(r_2)\phi_{2s}^*(r_1) \tag{13.21}$$

is called the exchange integral. These two integrals can be evaluated by employing the wavefunctions

$$\phi_{1s} = \frac{1}{\sqrt{\pi}} \left(\frac{Z}{a_0}\right)^{3/2} e^{-Zr/a_0}, \tag{13.22}$$

$$\phi_{2s} = \frac{1}{4\sqrt{2\pi}} \left(\frac{Z}{a_0}\right)^{3/2} \left(2 - \frac{Zr}{a_0}\right) e^{-Zr/2a_0}. \tag{13.23}$$

The results obtained are

$$J = 0.105 \frac{Z^2e^2}{a_0} \tag{13.24}$$

$$\text{and} \quad K = 0.011 \frac{Z^2e^2}{a_0}. \tag{13.25}$$

Since $K < J$, it follows from (13.18) and (3.19) that the space-symmetric ${}^1S_0$ (para) state has higher energy than that of the space-antisymmetric ${}^3S_1$ (ortho) state:

$$E_s = -2.037\, e^2/a_0, \quad E_a = -2.125\, e^2/a_0. \tag{13.26}$$

This is to be compared with the experimental result

$$E_s = -2.147\, e^2/a_0, \quad E_a = -2.176\, e^2/a_0. \tag{13.27}$$

Problems

1. Calculate the Coulomb and exchange integrals for the excited state of the helium atom in which one of the electrons is in $1s$ orbital and the second in the $2p$ orbital.

2. As a sequence to the problem 1 of chapter 12, obtain the time dependence of the first excited state of helium and the period of exchange oscillations.

3. Suppose the two electrons in the helium atom have a small mass difference $\Delta m \ll m$. This will cause a transition between the ortho and para helium states. Find $\Delta m/m$ such that one per hundred atoms would have made transition from the ortho to para state in 10^9 years. Neglect spin–orbit interaction.
   Hint: Take

$$H_0 = \frac{p_1^2}{2m} + \frac{p_2^2}{2m} + V(r_1) + V(r_2),$$

$$H = \frac{p_1^2}{2\left(m + \frac{\Delta m}{2}\right)} + \frac{p_2^2}{2\left(m - \frac{\Delta m}{2}\right)} + V(r_1) + V(r_2),$$

   so that $H_I = H - H_0 = \frac{-\Delta m}{4m^2}(p_1^2 - p_2^2)$.

   To calculate required transition rate, use time-dependent perturbation theory.

Bibliography

[1] W. Heisenberg, *Z. Physik*, **39**, 499 (1927).

[2] D.R. Hartree, *Proc. Cambridge Phil. Soc.*, **26**, 89 (1928).

[3] V. Fock, *Z. Physik*, **61**, 126 (1930).

[4] E.A. Hylleras, *Z. Physik*, **54**, 347 (1929).

[5] G. Kellner, *Z. Physik*, **44**, 91 (1927).

14 Many-Electron Atoms

14.1 Self-Consistent Field Method

In earlier chapters, while exact solution of Schrödinger equation for hydrogen atom has been obtained, for the helium atom a perturbative as well variational calculation was done. For atoms with more than two electrons, the problem of obtaining such solutions becomes very complex. Hartree developed a self-consistent method for such atoms in which individual electrons move in a central field $\Phi(r)$ produced by the average motion of the rest of the electrons in addition to the field of the nucleus. The former is obtained from the equation

$$\Phi(\vec{r}) = e \int d^3r' \frac{|\psi(\vec{r}')|^2}{|\vec{r} - \vec{r}'|}, \tag{14.1}$$

where $\psi(\vec{r})$ is the wavefunction of a single electron which satisfies the Schrödinger equation

$$\left[\frac{-\nabla^2}{2m} + \Phi(r) + V(\vec{r})\right] \psi(\vec{r}) = E\psi, \tag{14.2}$$

where $V(r)$ is the potential field of the nucleus. Thus while one needs the wavefunction $\psi(\vec{r})$ to obtain Φ, one requires to know Φ to obtain $\psi(\vec{r})$. Equations (14.1) and (14.2) are to be self-consistently solved.

The above two equations can be obtained from the many-particle Hamiltonian

$$H = \sum_i \left(\frac{p_i^2}{2m} + V(\vec{r}_i)\right) + \frac{1}{2} \sum_{j \neq i} V_{ij}(|\vec{r}_i - \vec{r}_j|), \tag{14.3}$$

where $V_{ij} = e^2/|\vec{r}_i - \vec{r}_j|$ and three and more particle interactions are not considered. In Hartree self-consistent method, this Hamiltonian is reduced to an effective Hamiltonian

$$H_{eff} = \sum_i \left[\frac{p_i^2}{2m} + V(\vec{r}_i) + e\Phi(\vec{r}_i)\right] = \sum_i H_i, \tag{14.4}$$

where

$$\Phi(\vec{r}_i) = \sum_{j \neq i} e \int d^3 r_j \frac{|\psi_j(\vec{r}_j)|^2}{|\vec{r}_i - \vec{r}_j|}, \tag{14.5}$$

and

$$H_i \psi_i(\vec{r}_i) = \frac{1}{2m} p_i^2 + V(\vec{r}_i) + e\Phi(\vec{r}_i) = E_i \psi_i(\vec{r}_i). \tag{14.6}$$

The above reduced or effective Hamiltonian is obtained by averaging of the two-particle interaction term in (14.3)

$$\Phi(\vec{r}_i) = \sum_{j \neq i} \langle \Psi_j | V_{ij} | \Psi_j \rangle, \tag{14.7}$$

where $|\Psi_j>$ is a single-electron eigenstate. Using the set of vector $|\vec{r}>$ with completeness relation

$$\int d^3 r |\vec{r}\rangle\langle\vec{r}| = 1 \tag{14.8}$$

in (14.7), we get

$$\Phi(\vec{r}_i) = \sum_{j \neq i} \int d^3 r' \langle \Psi_j | \vec{r}' \rangle \langle \vec{r}' | V_{ij} | \Psi_j \rangle \tag{14.9}$$

$$= e^2 \sum_{j \neq i} \int d^3 r'_j \frac{|\psi_j(r'_j)|^2}{|\vec{r}_i - \vec{r}'_j|}, \tag{14.10}$$

which is (14.5).

The two fundamental equations of the self-consistent field method are (14.5) and (14.6). To solve these coupled equations, one has to make an intelligent guess of the single-particle wavefunction $\psi(\vec{r})$ and obtain $\Phi(\vec{r})$ by making use of (14.5) and insert this $\Phi(\vec{r})$ in (14.6) to obtain $\psi(\vec{r})$ and then repeat this process several times till one recovers the initially guessed wavefunction. If one starts with $\phi_i(\vec{r}_i)$ as the single-particle wavefunction for the trial wavefunction,

$$E = \sum_i E_i = \sum_i \int \phi_i^* \left(\frac{p_i^2}{2m} + V(r_i) \right) \phi_i d^3 r$$

$$+ \frac{1}{2} \sum_{i,j \neq i} \int d^3 r_i d^3 r_j \phi_i^* \phi_j^* V_{ij} \phi_i \phi_j, \tag{14.11}$$

and demands that

$$\delta E_j = 0 \quad \text{for} \quad \int (\delta\phi_i^*) \phi_i d^3 r + c.c. = 0,$$

then from (14.11) one gets

$$\delta E = \sum_i \int \delta\phi_i^* \left\{ \left(\frac{p_i^2}{2m} + V(r_i) \right) + \sum_{k \neq i} \int \phi_k^* V_{ki} \phi_k d^3 r - E_i \right\}$$

$$\times \phi_i d^3 r_i + c.c. = 0. \tag{14.12}$$

Since the variations $\delta\phi_i^*$ and $\delta\phi_i$ are independent, this equation can be satisfied if

$$\left[\frac{p_i^2}{2m} + V(r_i) + \Phi_i - E_i \right] \phi_i = 0, \tag{14.13}$$

where

$$\Phi_i = \sum_{k \neq i} \int \phi_k^* V_{ki} \phi_k d^3 r_k. \tag{14.14}$$

Equation (14.13) was first proposed by Hartree in 1928 from physical considerations and later derived by Fock in 1930 using variational principle.

Let the wavefunction to start with be called the zeroth approximation

$$\phi_i(\vec{r}_i) = \phi_i^{(0)}(\vec{r}_i); \tag{14.15}$$

and use it in (14.14) to obtain

$$\Phi_i^{(0)} = \sum_{k \neq i} \int \phi_k^{(0)} V_{ki} \phi_k^{(0)} d^3 r_k, \tag{14.16}$$

and then use this in (14.13) to get

$$\left[\frac{p_i^2}{2m} + V(\vec{r}_i) + e\Phi_i^{(0)} - E_i^{(0)} \right] \phi_i^{(1)} = 0. \tag{14.17}$$

$\phi_i^{(1)}$ obtained from solution of this equation used in (14.14) gives

$$\Phi_i^{(1)} = \sum_{k \neq i} \int \phi_k^{(1)} V_{ki} \phi_k^{(1)} d^3 r_k. \tag{14.18}$$

This is used once again in (14.13) which now reads

$$\left[\frac{p_i^2}{2m} + V(r_i) + e\Phi_i^{(1)} - E_i^{(1)} \right] \phi_i^{(2)} = 0. \tag{14.19}$$

This process is repeated till one obtains equations (14.13) and (14.14) with the zeroth wavefunction. The final potential $\Phi_i(\vec{r}_i)$ is called the Hartree self-consistent field.

The above treatment does not use symmetrised wavefunctions as demanded by the statistics obeyed by the particles. This was done by Fock. We shall not go into the details of this calculation.

14.2 Statistical Method

Solution of the self-consistent field equations which is done numerically, becomes more and more difficult as one proceeds to heavier atoms with increasing number of electrons. For such atoms a statistical method developed by Thomas and Fermi is applied and very interesting conclusions about charge density and potential as a function of radius are obtained.

In order to calculate electron density, this method makes use of the fact that the majority of electrons in complex atoms have large principal quantum numbers as a result of which semiclassical physics can be used. Counting of number of electron states is done in terms of cells in phase space taking cell volume to be $(2\pi\hbar)^3$. According to Pauli principle, each cell is occupied by two electrons. Hence the number of electrons with momenta less than p will then be

$$n(r) = 2 \times \frac{4\pi}{3}\frac{p^3(r)}{(2\pi\hbar)^3} = \frac{p^3(r)}{3\pi^2\hbar^3}. \tag{14.20}$$

Further, the maximum total energy E_0 in a stationary state must be same at every point in space; otherwise electrons would move from one point to another. The maximum momentum p_0 would then be given by

$$\frac{p_0^2}{2\mu} - e\phi(r) = E_0, \tag{14.21}$$

where $\phi(r)$ is the electrostatic potential at a distance r from the nucleus.

From (14.20) and (14.21) we get

$$n(r) = \frac{[2\mu(e\phi(r) + E_0)]^{3/2}}{3\pi^2\hbar^3}. \tag{14.22}$$

Since at all points outside the atom, say at $r = R$, the electron density must vanish,

$$-e\phi(R) = E_0, \tag{14.23}$$

and since the atom as whole is neutral, the potential at all these points must vanish,

$$\phi(R) = 0. \tag{14.24}$$

Hence, according to (14.23)

$$E_0 = 0. \tag{14.25}$$

Substitution of this in (14.22) and using the resulting n in the Poisson's equation

$$\nabla^2\phi = 4\pi ne \tag{14.26}$$

gives

$$\nabla^2\phi = \frac{4e}{3\pi\hbar^3}[2\mu e\phi]^{3/2}. \tag{14.27}$$

This is the Thomas–Fermi equation which is to be solved with the boundary condition

$$\lim_{r\to 0}\phi(r) = \frac{Ze}{r} \tag{14.28}$$

since the potential is that of the nucleus alone, and

$$\lim_{r\to\infty} r\phi(r) = 0. \tag{14.29}$$

The Thomas–Fermi equation (14.27) and the boundary conditions (14.28) and (14.29) can be put in the dimensionless forms:

$$\sqrt{x}\frac{d^2f(x)}{dx^2} = f^{3/2}(x), \; f(0) = 1, \; f(\infty) = 0 \tag{14.30}$$

in terms of dimensionless variable x defined by

$$r = bxZ^{-1/3},$$

$$b = \frac{1}{2}\left(\frac{3\pi}{4}\right)^{2/3} a_0, \tag{14.31}$$

where

$$a_0 = \frac{\hbar^2}{\mu e^2}$$

and

$$\phi = \frac{Z^{4/3}}{bx}f(x). \tag{14.32}$$

Equation (14.30) can be solved numerically. The results give

$$f(\infty) = 0,$$

$$\lim_{x\to 0} f(x) = 1 - 0.59x, \tag{14.33}$$

which show that the atom extends to infinity. Substtution (14.33), (14.31) and (14.32) in (14.27) gives

$$\lim_{r\to 0}\phi(r) = \frac{Z}{r} - 1.80Z^{4/3}. \tag{14.34}$$

Here the first term is due to the nucleus and the second is due to all the electrons.

The electron density distribution $n(r)$ given in (14.22) can be written as

$$n(r) = Z^2 F\left(\frac{rZ^{1/3}}{b}\right), \tag{14.35}$$

where

$$F(x) = \frac{32}{9\pi^3}\left(\frac{f(x)}{x}\right)^{3/2},$$

which shows that the electron density distribution is of the same form in all atoms with $bZ^{-1/3}$ as a characteristic length and the maximum occurs at distances of the order $Z^{-1/3}$. Also, a numerical calculation shows that half the total number of electrons lie within a sphere of radius $1.33Z^{-1/3}$. Calculations for other parameters such as velocity distribution have been performed.

Bibliography

[1] D.R. Hartree, *Proc. Cambridge, Phil. Soc.* **24**, 111 (1928).
[2] V. Fock, *Zects f. Phys.*, **61**, 126 (1930).
[3] J.C. Slater, *Phys. Rev.*, **34**, 1293 (1929).
[4] L.H. Thomas, *Proc. Cambridge, Phil. Soc.*, **23**, 542 (1927).
[5] E. Fermi, *Z. Phys.*, **28**, 73 (1928).

15 Emission and Absorption of Photons by Atoms

15.1 Introduction

We shall take up the theory of the emission and absorption of photons by atoms as well as the photoelectric effect and radiative electron capture by proton by making use of the time-dependent perturbation theory developed in chapter 10. We shall restrict our treatment to the first-order perturbation. Under the action of the perturbing potential, the quantum systems undergo transition from an initial to a final state resulting either in emission or absorption of radiation. The probability of transition resulting in the emission of a single photon is given by the Fermi golden rule (10.85):

$$dW = \frac{2\pi}{\hbar}|V_{fi}|^2\delta(E_i - E_f)\rho(E_f)dE_f, \tag{15.1}$$

where $\rho(E_f)$ = density of final states. Here the initial state is the nth state of the atom and the final state is its mth state plus photon so that

$$E_i = E_n \quad \text{and} \quad E_f = E_m + \hbar\omega. \tag{15.2}$$

If the photon has a definite angular momentum,

$$dE_f = \hbar d\omega \quad \text{and} \quad \rho(E_f) = 1. \tag{15.3}$$

Substituting (15.2) and (15.3) in (15.1), we have

$$dW_{mn} = \frac{2\pi}{\hbar}|V_{mn}|^2\delta(E_n - E_m - \hbar\omega)d\omega \tag{15.4}$$

and

$$W_{mn} = \int dW_{mn} = \frac{2\pi}{\hbar}|V_{mn}|^2. \tag{15.5}$$

If on the other hand the photon has a definite momentum k,

$$dW = \frac{2\pi}{\hbar}|V_{mn}|^2\delta(E_n - E_m - \hbar\omega)\frac{d^3k}{(2\pi)^3} \tag{15.6}$$

where

$$d^3k = k^2 dk d\Omega_k = \omega^2 d\omega d\Omega_k.$$

Integration over ω gives

$$dW_{mn} = \frac{1}{4\pi 2\hbar}|V_{mn}|^2\omega_{nm}^2 d\Omega_k \tag{15.7}$$

with

$$\hbar\omega = E_n - E_m. \tag{15.8}$$

15.2 Spontaneous Emission of Photons of Definite Momentum

Under the action of the electromagnetic field, the Hamiltonian of the atom is given by

$$H = \frac{1}{2\mu}\left(\vec{p} - \frac{e\vec{A}}{c}\right)^2 + V_a = \frac{p^2}{2\mu} + V_a - \frac{e}{\mu c}\vec{A}\cdot\vec{p} + \frac{e^2A^2}{2\mu c^2}, \tag{15.9}$$

where $\vec{A}$ is the vector potential of the field satisfying $\vec{\nabla}\cdot\vec{A} = 0$. While the third term gives rise to emission, absorption and scattering of photons (in second order!) the last term gives rise to scattering only. For the emission of photon we shall therefore consider the first three terms:

$$H = H_a - \frac{e}{mc}\vec{A}\cdot\vec{p} = H_a + V, \tag{15.10}$$

with $H_a = \frac{p^2}{2m} + V_a$ and $V = \frac{e}{mc}\vec{A}\cdot\vec{p}$. It may be noted here that in spontaneous emission there is no real electromagnetic field to perturb the atom. The vacuum fluctuations of the field polarise the atom in the excited state as a result of which it emits a photon.

The matrix element of V is then

$$V_{mn} = -\frac{e}{mc}\int d^3r\vec{A}\cdot\psi_m^*\vec{p}\psi_n. \tag{15.11}$$

Taking

$$\vec{A} = \left(\frac{2\pi\hbar c^2}{\omega}\right)^{1/2} \vec{\mathcal{E}}_\omega e^{-i\vec{k}\cdot\vec{r}} \tag{15.12}$$

to be the matrix element of $\vec{A}$ between vacuum and the emitted photon, since the initial state is the excited atom in vacuum and the final state is the atom and the emitted photon as per (12.53), where $\vec{\mathcal{E}}_\omega$ is polarisation (unit) vector, we have

$$V_{mn} = -\frac{e}{m}\left(\frac{2\pi\hbar}{\omega}\right)^{1/2} \int d^3r e^{-i\vec{k}\cdot\vec{r}} \psi_m^*(\vec{r})\vec{p}\cdot\vec{\mathcal{E}}_\omega \psi_n(\vec{r}). \tag{15.13}$$

In order to evaluate the integral, we adopt the following approximation procedure. We expand the plane wave

$$e^{-i\vec{k}\vec{r}} = 1 - i\vec{k}\vec{r} + \cdots \tag{15.14}$$

since successive terms give decreasing contribution to the integral on account of exponential decrease in the radial dependence of the wavefunctions of the states of the atom. In other words,

$$\langle\vec{k}\cdot\vec{r}\rangle_{av} \sim ka = \frac{2\pi a}{\lambda} \ll 1 \tag{15.15}$$

for wavelength of the photon λ large compared to the dimension a of the atom. The first term in (15.14) when used in (15.13) gives what is known as the electric dipole or $E1$ radiation and the second term gives the electric quadrupole ($E2$) plus the magnetic dipole or $M1$ radiation. The corresponding transition matrix elements and probabilities are derived below.

Electric dipole ($E1$) radiation

Putting

$$e^{-i\vec{k}\cdot\vec{r}} = 1 \tag{15.16}$$

in (15.13) gives

$$V_{mn} = -\frac{e}{m}\left(\frac{2\pi\hbar}{\omega}\right)^{1/2} \vec{\mathcal{E}}_\omega \cdot \int d^3r \psi_m^*(\vec{r})\vec{p}\psi_n(\vec{r}). \tag{15.17}$$

Since

$$\vec{p} = m\frac{d\vec{r}}{dt} = -im[H_0, \vec{r}],$$

$$\psi_m^* \vec{p} \psi_n = im(E_n - E_m)\psi_m^* \vec{r} \psi_n \tag{15.18}$$
$$= i\frac{m}{e\hbar}\omega_{nm}\psi_m^* \vec{d} \psi_n,$$

where $\vec{d} = e\vec{r}$ is the electric dipole moment of the atom, we have

$$V_{mn} = \left(\frac{2\pi}{\hbar\omega}\right)^{1/2} \omega_{nm} \vec{d}_{mn} \cdot \vec{\mathcal{E}}_\omega. \tag{15.19}$$

It is clear why the radiation is called an electric dipole radiation.

Since $\vec{d}$ has odd parity, the states m and n must have opposite parity and since parity $P = (-1)^l$, the angular momentum quantum number of these states must differ by 1 unit, i.e. $|\Delta l| = 1$ and hence $\Delta m = 0, \pm 1$. Substitution of (15.19) in (15.7) gives

$$dW_{mn} = \frac{\omega^3}{2\pi\hbar c^3}|\vec{\mathcal{E}} \cdot \vec{d}_{mn}|^2 d\Omega_k. \tag{15.20}$$

Summation over polarisation $\vec{\mathcal{E}}$ gives

$$dW_{mn} = \frac{\omega^3}{2\pi\hbar c^3}|\vec{n} \times \vec{d}_{mn}|^2 d\Omega_k, \tag{15.21}$$

(where $\vec{n}$ = unit vector along $\vec{k}$) which on integration over all directions of the photon gives

$$W_{mn} = \frac{4\omega^3}{3\hbar c^3}|\vec{d}_{mn}|^2 \tag{15.22}$$

for the total probability of radiation. The intensity I of radiation is given by

$$I_{mn} = \hbar\omega W_{mn} = \frac{4\omega^4}{3c^3}|\vec{d}_{mn}|^2, \tag{15.23}$$

which is the same as the classical formula if $\vec{d}_{mn}$ is replaced by $\vec{d}$. The angular distribution of the electric dipole radiation can be calculated for the $\Delta m = 0$ and $\Delta m = \pm 1$ transitions from (15.21) as follows:

1. For $\Delta m = 0$, the matrix elements of d_1 and d_2 vanish and hence

   $$\delta W \sim |\vec{n} \times \vec{d}|^2 = d_3^2 \sin^2\theta, \tag{15.24}$$

   where θ = the angle between the direction of radiation with respect to that of the dipole.

2. For $\Delta m = \pm 1$, the matrix elements d_3 vanish and

$$|(\vec{n} \times \vec{d})|^2 = (d_1^2 + d_2^2)\cos^2\theta + (d_2\cos\phi - d_1\sin\phi)^2.$$

Averaging over ϕ gives

$$\overline{|\vec{n} \times \vec{d}|^2} = \frac{1}{2}(d_1^2 + d_2^2)(1 + \cos^2\theta). \tag{15.25}$$

Equations (15.24) and (15.25) give the angular distribution of the radiation.

It may be noted here that (15.22) gives the spontaneous radiation probability. The induced (stimulated) emission is obtained by multiplying it by a factor $\frac{\pi^2 c^2}{\hbar\omega^3} I$:

$$W_{mn(\text{stimulated})} = \frac{4\pi^2 I}{3\hbar^2 c}|\vec{d}_{mn}|^2, \tag{15.26}$$

where I is the intensity of the stimulating radiation.

Magnetic dipole ($M1$) and electric quodrupole (E_2) radiation

As we have stated, the contribution of the second term in (15.14) can be neglected when $\langle kr \rangle \ll 1$, i.e., λ is large compared to atomic dimension. For smaller λ, i.e. higher frequencies or for radiation from nuclei where dimensions are very small compared to those of atoms $\langle kr \rangle \sim \lambda$, the second term $i\vec{k} \cdot \vec{r}$ cannot be neglected. To obtain the matrix elements of $(\vec{k} \cdot \vec{r})p_i$ we proceed as follows:

$$\begin{aligned}\vec{k} \times \vec{L} &= \vec{k} \times (\vec{r} \times \vec{p}) \\ &= (\vec{k} \cdot \vec{p})\vec{r} - (\vec{k} \cdot \vec{r})\vec{p} = (\vec{k} \cdot \vec{p})\vec{r} + (\vec{k} \cdot \vec{r})\vec{p} - 2(\vec{k} \cdot \vec{r})\vec{p}\end{aligned}$$

Hence

$$\begin{aligned}(\vec{k} \cdot \vec{r})p_i &= \frac{1}{2}(\vec{L} \times \vec{k})_i + \frac{1}{2}mk_j(\dot{r}_j \cdot r_i + r_j \cdot \dot{r}_i) \\ &= \frac{1}{2}(\vec{L} \times \vec{k})_i + \frac{1}{2}mk_j\frac{d}{dt}(r_i r_j).\end{aligned} \tag{15.27}$$

In terms of the magnetic dipole moment vector $\vec{M}$,

$$\vec{M} = \frac{e\vec{L}}{2mc}, \tag{15.28}$$

and the electric quandrupole moment tensor Q_{ij},

$$Q_{ij} = er_i r_j, \tag{15.29}$$

(15.27) becomes

$$(\vec{k}\cdot\vec{r})p_i = \frac{mc}{e}(\vec{M}\times\vec{k})_i + \frac{1}{2e}mk_j\dot{Q}_{ij} \tag{15.30}$$

which on substitution in

$$V_{mn} = -\frac{e}{m}\left(\frac{2\pi\hbar}{\omega}\right)^{1/2}\int d^3r(-i\vec{k}\cdot\vec{r})\psi_m^*(\vec{p}\cdot\vec{\mathcal{E}})\psi_n \tag{15.31}$$

gives

$$V_{mn} = \frac{ie}{m}\left(\frac{2\pi\hbar}{\omega}\right)^{1/2}\int d^3r\psi_m^*[\vec{\mathcal{E}}\cdot(\vec{M}\times\vec{k}) + \frac{i}{2}\vec{\mathcal{E}}\cdot\vec{k}_j\omega_{mn}(\dot{Q}_{ij})_{mn}]. \tag{15.32}$$

The first term represents the magnetic dipole ($M1$) and the second the electric quadrupole ($E2$) matrix element. The calculation of the probability and the intensity of radiation is done as in the electric dipole case.

It may be noted that since

$$M \sim \frac{e\hbar}{mc},\quad d \sim ea = \frac{\hbar^2}{me}, \tag{15.33}$$

so that

$$\frac{M}{d} = \frac{e^2}{\hbar c} = \alpha, \tag{15.34}$$

the

$$\left(\frac{M1}{E1}\right)_{\text{intensity}} \sim \alpha^2 \tag{15.35}$$

for equal frequencies. Similar considerations give

$$\left(\frac{E2}{M1}\right)_{\text{intensity}} = \left(\frac{ma^2\omega}{\hbar}\right)^2 = \left(\frac{\Delta E}{E}\right)^2 \tag{15.36}$$

where $E \sim \hbar^2/ma^2 =$ the energy of the state and $\Delta E = \hbar\omega \sim$ the energy of the radiation.

Thus, for medium atomic frequencies for which $\Delta E \sim E$, $E2$ and $M1$ radiations are equally probable. But for transitions between fine structure and hyperfine structure components where $\Delta E \ll E$, $M1$ transition dominates over $E2$ transition. On the other hand, for radiation between nuclear states, $\Delta E > E$ so that $E2$ dominates over $M1$.

15.3 Spontaneous Emission of Photons of Definite Angular Momentum

In chapter 5, section 4, we have seen that photons with definite angular momentum are of two types — electric photons and magnetic photons whose wavefunctions have been derived. For the calculation of the transition amplitude for the radiation of these photons, we need to have expressions for the vector and scalar potentials $\vec{A}$ and Φ. Let us recall that for the electric photon

$$\vec{\psi}_{lm} = \vec{Y}_{lm}^{(e)}(\theta\varphi) = \frac{1}{\sqrt{l(l+1)}}\vec{\nabla}Y_{lm}(\theta\varphi). \tag{15.37}$$

Since $\vec{\psi} = \vec{E} + i\vec{B}$, it will be convenient to adopt a gauge where $\vec{A} = 0$, $\Phi \neq 0$ so that $\vec{B} = 0$.

$$\vec{\psi} = -\vec{\nabla}\Phi. \tag{15.38}$$

Comparison with (15.37) and after suitable normalisation gives

$$\Phi_{lm}(k) = -\sqrt{\frac{l+1}{l}}\frac{4\pi^2}{\omega^{3/2}}Y_{lm}(\theta\varphi)\delta(k-\omega). \tag{15.39}$$

Similarly, for the magnetic photon

$$\vec{\psi}_{lm} = \vec{Y}_{lm}^{(m)}(\theta,\varphi) = \vec{n} \times \vec{Y}_{lm}^{(e)}(\theta\varphi), \tag{15.40}$$

we can adopt the gauge $\Phi = 0$, $\vec{A} \neq 0$, with

$$\vec{A}_{lm}(\vec{k}) = \frac{4\pi^2}{\omega^{3/2}}\delta(k-\omega)\vec{Y}_{lm}^{(m)}(\hat{k}). \tag{15.41}$$

With these preliminaries, we can now work out the transition matrix elements, emission probabilities and angular distributions of the electric and magnetic photons.

Electric photon

The Schrödinger equation with $\vec{A} = 0$, $\Phi \neq 0$ becomes

$$\left(\frac{p^2}{2m} + V_a\right)\psi = (H_0 - e\Phi(r))\psi = E\psi, \tag{15.42}$$

so that the interaction of the atom with the photon is given by

$$V = -e\Phi(r) = -e\int d^3k/(2\pi)^3 e^{i\vec{k}\cdot\vec{r}}\Phi(\vec{k}) = \sqrt{\frac{l+1}{l}}\int \frac{4\pi^2}{\omega^{3/2}}\delta(k-\omega)$$

$$\times\, Y_{lm}(\hat{k})\frac{d^3k}{(2\pi)^3}e^{-ik\vec{r}} \tag{15.43}$$

and the matrix element becomes

$$V_{mn} = -e\int d^3r\psi_m^*(\vec{r})\Phi(r)\psi_n(r)$$

$$= \frac{-e\sqrt{\omega}}{2\pi}\sqrt{\frac{j+1}{j}}\int d^3r\rho_{mn}(\vec{r})\int d\Omega_k e^{-i\vec{k}\cdot\vec{r}}Y_{lm}(\hat{k}), \tag{15.44}$$

where

$$\rho_{mn}(\vec{r}) = \psi_m^*(\vec{r})\psi_n(\vec{r}).$$

Substituting the expansion formula

$$e^{-i\vec{k}\cdot\vec{r}} = 4\pi\sum_{l=o}^{\infty}\sum_{m=-l}^{l} i^l g_l(kr)Y_{lm}^*(\hat{k})Y_{lm}(\hat{r}) \tag{15.45}$$

where

$$g_l(kr) = \sqrt{\frac{\pi}{2kr}}\, J_{l+\frac{1}{2}}(kr) \tag{15.46}$$

is the spherical Bessel function, into (15.44) and performing the $d\Omega_k$ integration gives

$$\int d\Omega_k e^{-i\vec{k}\cdot\vec{r}}Y_{lm}(\hat{k}) = 4\pi(i)^{-j}g_l(kr)Y_{lm}(\hat{r}). \tag{15.47}$$

For radiation wavelengths with $a/\lambda \ll 1$, where a is of atomic dimension, only r values in the d^3r integration in (15.44) will be important for which $kr \ll 1$, so that we can approximate $g_l(kr)$ to

$$g_l(kr) \approx (kr)^l/(2l+1)!!. \tag{15.48}$$

Inserting (15.47) and (15.48) into (15.44) gives

$$V_{mn} = (-1)^{m+1}i^l\sqrt{\frac{(2l+1)(l+1)}{l\pi}}\;\frac{\omega^{l+\frac{1}{2}}}{(2l+1)!!}\,e(Q_{l,m}^{(e)})_{mn}, \tag{15.49}$$

where

$$\left(Q_{lm}^{(e)}\right)_{mn} = \sqrt{\frac{4\pi}{2l+1}} \int d^3r \rho_{mn}(\vec{r}) r^l Y_{lm}(\hat{r}) \tag{15.50}$$

is the 2^l pole electric transition moment of the atom. If we make use of $\rho_{mn} = \psi_m^* \psi_n$, we find that

$$Q_{lm}^{(e)} = \sqrt{\frac{4\pi}{2l+1}}\, r^l Y_{lm} \tag{15.51}$$

is the 2^l pole electric moment operator. To obtain the probability of E_l electric photon radiation, we substitute (15.49) into (15.5) and obtain

$$W_{lm}^{(e)} = \frac{2(2l+1)(l+1)e^2}{l[(2l+1)!!]^2} \omega^{2l+1} |(Q_{l,-m}^{(e)})_{mn}|^2. \tag{15.52}$$

For determining the angular distribution of the electric photon we make use of its wavefunction

$$\vec{\psi} = -i\vec{\nabla}\Phi,$$

whose angular part is

$$\vec{\psi}_{lm} \sim \vec{\nabla} Y_{lm}(\vec{k}), \tag{15.53}$$

so that

$$dW_{lm} \sim |\vec{\nabla}_k Y_{lm}(\vec{k})|^2 d\Omega_k. \tag{15.54}$$

For $l = 1$, this gives for $m = 1, 0, -1$

$$dW_{1,0} \sim \sin^2\theta d\Omega_k,$$

$$dW_{1,\pm 1} \sim \frac{1+\cos^2\theta}{2} d\Omega_k, \tag{15.55}$$

which is same as (15.24) and (15.25). The order of the magnitude estimates show that the electric photon probability goes as

$$W_{lm}^{(e)} \sim \alpha k (ka)^{2l}, \tag{15.56}$$

where a is of atomic dimensions. Since $ka \ll 1$, this probability decreases with l.

Magnetic photon

As already stated, for a magnetic photon emission, the gauge with

$$\vec{A}_{lm} = \frac{4\pi^2}{\omega^{3/2}}\delta(k-\omega)\vec{Y}^{(m)}_{lm}(\hat{k}),\ \ \Phi = 0 \tag{15.57}$$

is found convenient. The Schrödinger equation

$$\left[\frac{1}{2m}(\vec{p}-e\vec{A})^2 + V_a\right]\psi = (H_0 + V)\psi = E\psi, \tag{15.58}$$

gives

$$V = e\frac{\vec{A}\cdot\vec{p}}{m} = \vec{J}\cdot\vec{A}, \tag{15.59}$$

where $\vec{J}$ = current density. It's matrix element works out to

$$V_{mn} = \frac{-e}{2\pi}\omega^{\frac{1}{2}}\int d^3r\,\vec{J}_{mn}\int \vec{Y}^{(m)}_{lm}(k)e^{-i\vec{k}\cdot\vec{r}}d\Omega_k \tag{15.60}$$

which simplifies to

$$V_{mn} = -ei^{-l}\frac{2\omega^{l+\frac{1}{2}}}{(2l+1)!!}\int d^3rr^j\vec{J}_{mn}\vec{Y}^{(m)}_{lm}(\hat{r}) \tag{15.61}$$

by making use of the integral

$$\int d\Omega_k e^{-i\vec{k}\cdot\vec{r}}\vec{Y}^{(m)}_{lm}(\hat{k}) = 4\pi i^{-l}g_j(kr)\vec{Y}^{(m)}_{lm}(\hat{r}). \tag{15.62}$$

Further simplification is achieved by using the relation

$$\vec{Y}^{(m)}_{lm}(\hat{r}) = \frac{1}{\sqrt{l(l+1)}}\vec{r}\times\vec{\nabla}Y_{lm}, \tag{15.63}$$

which yields

$$V_{mn} = (-1)^m i^l\left[\frac{(2l+1)(l+1)}{l}\right]^{1/2}\frac{\omega^{l+\frac{1}{2}}}{(2l+1)!!}e(Q^{(m)}_{l,-m})_{mn}, \tag{15.64}$$

where

$$(Q^{(m)}_{lm})_{mn} = \frac{1}{l+1}\left(\frac{4\pi}{2l+1}\right)^{1/2}\int d^3r\vec{r}\times\vec{J}_{mn})\vec{\nabla}(r^lY_{lm}) \tag{15.65}$$

is the 2^l pole magnetic transition moment. This matrix element for magnetic photon emission is the same as that for the electric photon emission obtained in the previous section except that it has magnetic transition moment in place electric transition moment. The total transition probability and angular distribution will also be similarly related. For $l = 1$, the magnetic transition moment becomes

$$(Q_{1m}^{(m)})_{mn} = \frac{1}{2}\int d^3r\vec{r} \times \vec{J}_{mn}(\vec{r}), \tag{15.66}$$

which is the matrix element of the magnetic dipole moment operator.

15.4 Photoelectric Effect

When a photon is absorbed by an atom, it causes transition from the ground state of the atom to higher discrete as well as continuum states. In the latter case, the phenomenon is called photoelectric effect. The matrix element for this transition to the continuum from the ground state g can be written as

$$V_{cg} = -ie\int d^3r\frac{e^{-i\vec{k}\cdot\vec{r}}\vec{\mathcal{E}}_k}{\sqrt{2\omega}}\psi_c^*\vec{\nabla}\psi_g, \tag{15.67}$$

where we have taken

$$\vec{A} = \frac{\vec{\mathcal{E}}e^{-i\vec{k}\cdot\vec{r}}}{\sqrt{2\omega}}. \tag{15.68}$$

We shall confine ourselves to the dipole approximation:

$$e^{i\vec{k}\cdot\vec{r}} = 1. \tag{15.69}$$

Consider the photo-effect in hydrogen atom for which

$$\psi_g = \frac{1}{\sqrt{\pi a_0^3}}e^{-r/a_0}, \quad a_0 = \frac{\hbar^2}{me^2}. \tag{15.70}$$

The selection rule $\Delta l = 1$ for the dipole approximation requires that we take only the p-wave ($l = 1$) part of the continuum

$$\psi_c(kr) = \sqrt{\pi/2}\sum_l^{\infty} i^l(2l+1)R_{kl}(r)P_l(\hat{k}\cdot\hat{r}), \tag{15.71}$$

i.e.,

$$\psi_c = \frac{3}{k}\sqrt{\frac{\pi}{2}}R_{k1}(r)(\hat{k}\cdot\hat{r}), \tag{15.72}$$

where, $R_{k1}(r)$ is given by

$$R_{k1} = \frac{2e^2m}{3}\sqrt{\frac{1+\nu^2}{\nu(1-e^{-2\pi\nu})}}kre^{-ikr}F(2+i\nu, 4, 2ikr) \tag{15.73}$$

with $\nu = e^2/\hbar v$, v = velocity of electron. Substituting (15.70) and (15.72) in (15.67) and then using (15.69), we get

$$V_{cg} = 2^{7/2}\pi\nu^3\hat{k}k(1+\nu^2)^{3/2}\frac{\exp[-2\nu\cot^{-1}\nu]}{\sqrt{1-\exp(-2\pi\nu)}}. \tag{15.74}$$

This is to be substituted in the 'Fermi golden rule' formula for the cross-section

$$\begin{aligned} d\sigma &= 2\pi|V_{cg}|^2\delta(\omega - \frac{me^4}{2} - \epsilon)\frac{d^3k}{(2\pi)^3} \\ &= e^2\frac{p}{2\pi\omega m}|\vec{\mathcal{E}}\cdot\vec{p}_{cg}|^2 d\Omega_k \end{aligned} \tag{15.75}$$

with $\frac{me^4}{2}$ = the ionisation potential, i.e. the binding energy, and ϵ = the kinetic energy of the ejected electron. On doing so and integrating over k (p = momentum of relative motion), we get

$$d\sigma = 2^7\pi\alpha a_0^2\left(\frac{me^4}{2\omega}\right)^4\frac{\exp[-4\nu\cot^{-1}\nu]}{1-\exp(-2\pi\nu)}(\hat{k}\cdot\hat{\mathcal{E}})^2 d\Omega_k. \tag{15.76}$$

The total cross-section obtained after integrating over all angles, is

$$\sigma = \frac{2^9\pi^2\alpha}{3}a_0^2\left(\frac{me^4}{2\omega}\right)^4\frac{\exp[-4\nu\cot^{-1}\nu]}{1-\exp(-2\pi\nu)}. \tag{15.77}$$

The threshold cross-section is obtained by putting $\omega = \frac{me^4}{2}$.

$$\sigma_{th} = \frac{2^9\pi^2\alpha}{3e^4}a_0^2. \tag{15.78}$$

15.5 Radiative Electron Capture by Proton

A free electron cannot be captured by a free proton to form a hydrogen atom because of the lack of simultaneous conservation of energy and momentum. However this conservation holds if a photon is emitted in the process.

$$p + e \longrightarrow H + \gamma \tag{15.79}$$

This is called radiative electron proton capture. The matrix elements V_{fi} of the process can be obtained from the photoelectric effect by making use of the 'reciprocity theorem'

$$V_{fi} = V_{i^*f^*}, \tag{15.80}$$

where states $|i^* >$ and $|f^* >$ are obtained from $|i >$ and $|f >$ by changing the signs momenta and spin projections of the states, i.e., by time-reversing states $|i >$ and $|f >$. From (15.80) we get

$$\frac{1}{p_f^2}\frac{d\sigma_{fi}}{d\Omega_f} = \frac{1}{p_i^2} d\sigma_{i^* f^*} d\Omega_{i^*}, \tag{15.81}$$

which expresses what is known as the 'principle of detailed balancing'. From this we get the relation

$$g_i p_i^2 \sigma_{fi} = g_f p_f^2 \sigma_{if}, \tag{15.82}$$

where

$$g_{i,f} = (2S_f + 1) \tag{15.83}$$

are the statistical weights of the states. When we apply (15.82) to the radiative capture (recombination) problem, we obtain

$$\sigma_{\text{capture}} = \sigma_{\text{photo}} \frac{2k^2}{p^2} \tag{15.84}$$

since statistical weight of the photon = 2 and that of a spinless electron is unity. Here p = momentum of the electron and k that of the photon.

15.6 Stimulated (Induced) Emission and Absorption

A stimulated emission is the emission of photons induced by light, i.e., an assembly of real photons. As mentioned earlier, spontaneous emission takes place without such induction but through the interaction of the atom with vacuum fluctuations which can be termed as virtual photons. If there are n real photons in the inducing light, after emission of one photon there will be $(n+1)$ photons. In spontaneous emission there are no real photons in the inducing electromagnetic field and one photon is emitted. It is clear that the intensity for emission can be obtained by taking the matrix elements $\vec{A}$ between the n photon state and $(n+1)$ photon state, which means that one has to multiply (15.12) by $\sqrt{n+1}$ (as per (12.53)) and therefore,

$$I_{st} = (n+1) I_s, \tag{15.85}$$

where I_{st} is intensity of stimulated emission, I_s is the intensity of spontaneous emission and n is the number of stimulating photons.

Similarly, absorption of light consists of n photons in the initial state and $(n-1)$ photons in the final state. One has, therefore, to take the matrix element of $\vec{A}$ between the state of n photons and $(n-1)$ photons. This is proportional to $\sqrt{n}$. The intensity of absorption I_a will therefore be related to (15.12) by

$$I_a = n I_s, \tag{15.86}$$

where I_a is intensity of absorption from an assembly of n photons.

15.7 Thermodynamic Equilibrium between Atom and Radiation

We shall consider only two levels, the ground and the first excited state of the atom and the light (radiation) emitted by it. Labelling the excited state as m and ground state as n, the equilibrium condition can be written as

$$N_m(A_{mn} + \rho B_{mn}) = \rho N_n B_{nm}, \tag{15.87}$$

where A_{mn} B_{mn} are transition probabilities for spontaneous and induced emission respectively from the excited (m) to the ground state (n) and B_{nm} is the transition probability of absorption from the ground to the excited state. In term of

$$N_n = ce^{-\beta E_n}, \quad \beta = \frac{1}{kT}, \quad E_m - E_n = \hbar\omega, \tag{15.88}$$

(15.87) becomes

$$\rho(\omega) = \frac{A_{mn}/B_{mn}}{(B_{nm}/B_{mn})(e^{\beta\hbar\omega} - 1)}, \tag{15.89}$$

which becomes Planck's formula

$$\rho(\omega) = \frac{\hbar\omega^3}{\pi^3 c^3} \frac{1}{(e^{\beta\hbar\omega} - 1)} \tag{15.90}$$

if

$$B_{mn} = B_{nm} = \frac{\pi^2 c^3}{\hbar\omega^3} A_{mn}. \tag{15.91}$$

Substituting (15.22) for $A_{mn} = W_{mn}$, (15.91) gives

$$B_{mn} = \frac{4\pi^2}{3\hbar^2} |\vec{d}mn|^2, \tag{15.92}$$

$$A_{mn} = \frac{4\omega^3}{3\hbar c^3} |\vec{d}mn|^2. \tag{15.93}$$

The A_{mn} and B_{mn} are called Einstein coefficients, named after Einstein who first investigated this problem.

Problems

1. Calculate the intensity of dipole radiation from all the levels of $n = 2$ state to ground state of hydrogen atom which is under the influence of an external uniform electric field.

2. Calculate the intensity of dipole radiation transition from state n to s in a harmonic oscillator.

3. A state with zero angular momentum (state) makes a gamma transition to a lower state with unit angular momentum (P state). This state in turn makes a transition to yet another lower state with zero angular momentum (S state) after a very short time, emitting a second ray. Both the gamma quanta are observed with suitable detector. Obtain the probability of the directions of propagation of the two quanta making an angle θ.
Hint: Use the transition probability formula for gamma:

$$P(\theta) = \begin{cases} c(1 - \cos^2\theta) & \text{for } \Delta m = 0, \\ \frac{1}{2}c(1 + \cos^2\theta) & \text{for } \Delta m = \pm 1, \end{cases}$$

where θ = angle between the dipole and momentum vector of the gamma. Note that selection rule for dipole transition is $\Delta m = \pm 1$. Hence for γ_2,

$$P(\theta) = \frac{1}{2}c(1 + \cos^2\theta),$$

where θ is angle between momenta of γ_1 and γ_2. Obtain c from $\int P(\theta)\sin\theta d\theta d\varphi = 1$.

Bibliography

[1] A.A. Sokolv, I.M. Ternov and V. Ch. Zhukovskii, *Quantum Mechanics*, Mir Publishers, Moscow (1984).

[2] V.B. Berestetskii et al. *Relativistic Quantum Theory*, vol. 4 of the *Course of Theorectical Physics*, Pergamon Press, Oxford (1971).

16 Scattering of Photons by Atoms

16.1 Introduction

Scattering of photons by free and bound electrons takes place through the latter's interaction with the photon field. The relevant Hamiltonian is given by

$$H = \frac{1}{2m}(\vec{p} - e\vec{A})^2 + V_a = \frac{p^2}{2m} + V_a - \frac{e}{m}\vec{A}\cdot\vec{p} + \frac{e^2A^2}{2m}$$

$$= H_a - \frac{e\vec{A}\cdot\vec{p}}{m} + \frac{e^2A^2}{2m}, \tag{16.1}$$

where $H_a = \frac{p^2}{2m} + V_a$ describes the atom and as already mentioned in the previous chapter, the two terms containing vector potential cause scattering of photon by the atom. Scattering is a two-step process—the incident photon causes electrons to get excited, which on de-excitation emits photon with either the same energy but different momentum, or with different energy as well as momentum. The former is called elastic scattering and the latter inelastic scattering. This process is caused by the term $\frac{e\vec{A}\cdot\vec{p}}{m}$ containing both photon field $\vec{A}$ and electron momentum $\vec{p}$. One uses time-dependant perturbation theory, which is second order in this term. The term $\frac{e^2A^2}{2m}$ also gives rise to scattering for which the first-order perturbation theory is of the same order of smallness as the second-order theory for $\frac{e\vec{A}\cdot\vec{p}}{m}$ term.

Scattering of photon by a recoil-less heavy unbound charged particle will be treated in section 2 using first-order perturbation in $\frac{e^2A^2}{2m}$ term. This will give rise to the well known Thomson scattering. Scattering by electrons bound in atoms will be treated in section 3 using second-order perturbation in the $\frac{\vec{A}\cdot\vec{p}}{m}$ term. This will result in the Kramers–Heisenberg dispersion formula from which the inelastic Raman scattering will follow; this will be taken up in the same section. Instead of taking real photons, if one considers the related process whereby the atom emits a virtual photon ($\omega^2 - k^2 \neq 0$) which is subsequently absorbed, one obtains the line shift (Lamb shift) and line width as its real and imaginary parts. Details of this will be given in the next chapter.

16.2 Scattering of Photon by Charged Particle

In this section we consider the scattering of photon by recoil-less charged particle. The initial and final states in the scattering of photon by an infinitely heavy recoil-less electron are

$$|i\rangle = |\vec{k}, \vec{\epsilon}_{\vec{k}}\rangle, \qquad |f\rangle = |\vec{k}', \vec{\epsilon}_{\vec{k}'}\rangle, \tag{16.2}$$

where $\vec{k}$ and $\vec{k}'$ are the momenta of the incident and the scattered photon and $\vec{\epsilon}$ and $\vec{\epsilon}_{k'}$ are their polarisation vectors. Since the electron does not recoil, $\omega_k = \omega_{k'}$. The perturbating potential that causes the scattering is

$$V^{(s)} = \frac{e^2 \vec{A} \cdot \vec{A}}{2m}, \qquad c = 1. \tag{16.3}$$

The matrix element of this potential between initial and final states is

$$V_{fi}^{(s)} = \frac{e^2}{2m\omega_k} \vec{\epsilon}_{\vec{k}'} \vec{\epsilon}_{\vec{k}'} e^{i(\vec{k}-\vec{k}')\vec{x}}, \tag{16.4}$$

where we have taken

$$\vec{A}_{\vec{k}}(x) = \frac{\epsilon_{\vec{k}} e^{i\vec{k}\cdot\vec{x} - i\omega t}}{\sqrt{2\omega_k}} + c.c. \tag{16.5}$$

Making use of the Fermi golden rule, we have, for the transition probability per unit time,

$$dW_{fi} = \frac{2\pi}{\hbar} |V_{fi}|^2 \rho_f = \frac{2\pi}{\hbar} |V_{fi}|^2 \frac{\vec{k}'^2 d\Omega_{\vec{k}'}}{(2\pi)^3}. \tag{16.6}$$

We have to sum over all final states to get the cross-section for which we make use of the identity

$$\sum_{\vec{\epsilon}_{k'}} \vec{\epsilon}_{\vec{k}} \cdot \vec{\epsilon}_{\vec{k}'} = 1 - \frac{(\vec{k}' \cdot \epsilon_{\vec{k}})^2}{(\vec{k}')^2} \tag{16.7}$$

so that

$$\sum_{\vec{\epsilon}_{\vec{k}}} \sum_{\vec{\epsilon}_{\vec{k}'}} \vec{\epsilon}_{\vec{k}'} \cdot \vec{\epsilon}_{\vec{k}'} = \frac{1}{2}(1 + \cos^2\theta). \tag{16.8}$$

The differential and total cross-sections so obtained are

$$\frac{d\sigma}{d\Omega} = \frac{r_0^2}{2}(1 + \cos^2\theta), \quad \sigma = \frac{8\pi}{3} r_0^2 = 6.65 \times 10^{-25}\text{cm}^2, \tag{16.9}$$

where

$$r_0 = \frac{e^2}{4\pi m} \tag{16.10}$$

is the classical radius of the charged particle. Formula (16.9) was obtained by J.J. Thomson

16.3 Scattering of Photon by Bound Electron

In the previous section, the interaction term giving rise to scattering of photon by free electron did not contain any charged particle operators and was suited to the problem where the charged particle, which was considered infinitely heavy, did not change its energy; the photon momentum changed its direction only. For the case of an electron bound to the nucleus, both its energy and momentum can change. Therefore the interaction term $e\vec{A} \cdot \vec{p}/m$ has to be used in the second order of perturbation, the matrix element V_{fi} for which is,

$$\begin{aligned} V_{fi} &= \frac{e^2}{m^2} \sum_n \frac{A_j(\vec{k}')\langle f|p'_j|n > \langle n|p_l|i > A_l(\vec{k})}{E_i + \omega_k - E_n} \\ &+ \frac{e^2}{m^2} \sum_n \frac{A_j(\vec{k})\langle f|p_j|n\rangle\langle n|p'_l|i\rangle A_l(\vec{k}')}{E_i - \omega_{k'} - E_n} \\ &= \frac{e^2}{2m^2} \frac{1}{\sqrt{\omega_k \omega_{k'}}} \\ &\times \sum_n \left[\frac{(\vec{\epsilon}(\vec{k}') \cdot \vec{p}'_{fn})(\vec{\epsilon}(\vec{k}) \cdot \vec{p}_{ni})}{E_i + \omega_k - E_n} + \frac{(\vec{\epsilon}(\vec{k}) \cdot \vec{p}_{fn})(\vec{\epsilon}(\vec{k}') \cdot \vec{p}'_{ni})}{E_i - \omega'_k - E_n} \right]. \end{aligned} \tag{16.11}$$

This identity can be brought to a convenient form by employing the following kinematics:

$$\begin{aligned} \vec{p} &= m\dot{\vec{x}} = im[H, \vec{x}], \\ \vec{p}_{fn} &= im\omega_{fn}\vec{x}_{fm}. \end{aligned} \tag{16.12}$$

Further, using

$$[x_i, p_j] = i\delta_{ij}, \quad [x_i, x_j] = 0, \tag{16.13}$$

we have

$$\left(\vec{\epsilon}(\vec{k}') \cdot \vec{x}\right)\left(\vec{\epsilon}(\vec{k}) \cdot \vec{p}\right) - \left(\vec{\epsilon}(\vec{k}) \cdot \vec{p}\right)\left(\vec{\epsilon}(\vec{k}') \cdot \vec{x}\right) = i\vec{\epsilon}_{k'} \cdot \vec{\epsilon}_k,$$

$$\left(\vec{\epsilon}(\vec{k}')\cdot\vec{x}\right)\left(\vec{\epsilon}(\vec{k})\cdot\vec{x}\right)-\left(\vec{\epsilon}(\vec{k})\cdot\vec{x}\right)\left(\vec{\epsilon}(\vec{k}')\cdot\vec{x}\right)=0, \tag{16.14}$$

$$E_f - E_i = \omega_k - \omega_{k'}. \tag{16.15}$$

Using (16.12), (16.13), (16.14) and (16.15) we finally get

$$V_{fi} = \frac{e^2}{2}\sqrt{\omega_k \omega_{k'}} \times \sum_n \frac{\left(\vec{\epsilon}(\vec{k}')\cdot\vec{x}_{fn}\right)\left(\vec{\epsilon}(\vec{k})\cdot\vec{x}_{ni}\right)}{E_i+\omega_k-E_n} + \frac{\left(\vec{\epsilon}(\vec{k})\cdot\vec{x}_{fn}\right)\left(\vec{\epsilon}(\vec{k}')\cdot\vec{x}_{ni}\right)}{E_i-\omega_k'-E_n} \tag{16.16}$$

and

$$\frac{dW_{fi}}{d\Omega} = \left(\frac{e^2}{4\pi}\right)^2 \omega_k \omega_{k'}^3 \left| \sum_n \frac{\left(\vec{\epsilon}(\vec{k}')\cdot\vec{x}_{fn}\right)\left(\vec{\epsilon}(\vec{k})\cdot\vec{x}_{ni}\right)}{E_i+\omega_k-E_n} + \frac{\left(\vec{\epsilon}(\vec{k})\cdot\vec{x}_{fn}\right)\left(\vec{\epsilon}(\vec{k})\cdot\vec{x}_{ni}\right)}{E_i-\omega_{k'}-E_n} \right|. \tag{16.17}$$

This is the Kramer's–Heisenberg (dispersion) formula. It is to be noted that the energy denominators in both the terms in this formula can be written as $(E_f - E_n)$, where in the first term $E_f = E_i + \omega_k$ and in the second term $E_f = E_i - \omega_k'$. Therefore, when $\omega_k = \omega_{k'}$, $E_f = E_i$, i.e. the initial and the final states of the atom are the same. This elastic scattering is called Rayligh scattering. When $\omega_k \neq \omega_{k'}$, there arise two cases: (a) when $\omega_{k'} > \omega_k$ and (b) when $\omega_k > \omega_{k'}$. In the former, the scattered photon has greater energy than the incident photon which means that the photon has gained energy of the amount $(\omega_{k'} - \omega_k) > 0$, which in turn means $E_f < E_i$ because the total energy of the atom and the photon must be conserved in the scattering process. This is called Raman scattering. In the latter case, when $E_f > E_i$, the atom has gained energy and is called Stokes scattering, which happens normally. The former unusual case was observed for the first time by C.V. Raman in liquids and by Landsberg and Mandelshtam in crystals.

Bibliography

[1] H.A. Kramers and W. Heisenberg, *Zs. f. Phys.*, **31**, 681 (1925); *Nature* **114**, 2310 (1924).

[2] J.J. Thomson, *Conduction of Electricity through Gases*, 3rd edition, Cambridge University Press, Vol. II p. 256 (1933).

[3] C.V. Raman, *Ind. Journ. Phys.*, **2**, 387 (1928); G.S. Landsberg and L.I. Mandelshtam, *Naturwiss*, **16**, 557 (1928).

17 Lamb Shift

17.1 Introduction

The interaction of the electromagnetic field with an atom gives rise to correction to the energy levels calculated on the basis of Coulomb interaction between the nucleus and the electron. In the hydrogen atom there exists l-degeneracy. For example, $2S$ and $2P$ states have the same energy. Due to this correction, these two states will have different energies. This is known as the level shift, which was experimentally observed and measured by Lamb and Rutherford and called the Lamb shift. This was first calculated by Bethe using non-relativistic theory and Welton by considering the interaction of vacuum fluctuation of the electromagnetic field with the electron. The atomic electron emits a virtual photon $(k^2 - \omega^2) \neq 0$ which is subsequently absorbed by it. This gives rise to the shift.

17.2 Bethe Method

In the scattering of photon from the atom, we have an incident real photon which is absorbed and a scattered photon which is emitted. In the present case we have a virtual photon which is emitted and subsequently absorbed. In scattering, the photon is first absorbed and then emitted. Thus there exists a relation between the two problems. To obtain the level shift from the scattering amplitude, all that we have to do is to set the frequency and the momenta of the incident and the scattered photon as well those of the electron equal and integrate over the photon momentum. After this is done (as in (16.11) of the previous chapter), we get for the energy shift for the state $|0\rangle$ (not necessarily ground state)

$$\langle 0|V|0\rangle = \frac{e^2}{16\pi^3 m^2} \int \frac{d^3k}{\omega_k} \sum_n \frac{|\vec{\epsilon}_\lambda(\vec{k}) \cdot \vec{p}_{\text{on}}|^2}{E_0 - E_n - \omega_k - i\epsilon}, \tag{17.1}$$

where, as usual, the $i\epsilon$ in the denominator is added to the define the singularity and λ stands for polarisation of photon. From this, we get

$$\text{Re}\langle 0|V|0\rangle = \Delta E = \frac{e^2}{16\pi^3 m^2} P \int \frac{d^3k}{\omega_k} \sum_{n\lambda} \frac{|\vec{\epsilon}_\lambda(\vec{k}) \cdot \vec{p}_{\text{on}}|^2}{E_0 - E_n - \omega_k}, \tag{17.2}$$

$$\mathrm{Im}\langle 0|V|0\rangle = \Gamma = \frac{e^2}{8\pi^2 m^2}\int \frac{d^3k}{\omega_k}\sum_{n\lambda}|\vec{\epsilon}_\lambda(\vec{k})\cdot\vec{p}_{\mathrm{on}}|^2\delta(E_0 - E_n - \omega_k), \tag{17.3}$$

where P stands for principal value. Thus,

$$\langle 0|V|0\rangle = \left(\Delta E - \frac{i\Gamma}{2}\right), \tag{17.4}$$

so that the wavefunction after this correction becomes

$$\psi(\vec{x}t) = \psi(\vec{x})e^{-iE-i\Delta Et-\frac{1}{2}\Gamma t}, \tag{17.5}$$

which shows that Γ is the level width. After some elementary calculations (17.2) becomes

$$\Delta E = \frac{e^2}{6\pi^2 m^2}P\int_0^\infty d\omega_k\omega_k\sum_n\frac{|\vec{p}_{\mathrm{on}}|^2}{E_0 - E_n + \omega_k}, \tag{17.6}$$

which is linearly divergent. We split this into two parts by using the identity

$$\frac{\omega_k}{E_0 - E_n - \omega_k} = \frac{E_0 - E_n}{(E_0 - E_n - \omega_k)} - 1, \tag{17.7}$$

$$\Delta E = \Delta E_B + \Delta E_F, \tag{17.8}$$

where

$$\Delta E_B = \frac{e^2}{6\pi^2 m^2}P\int_0^\infty d\omega_k\sum_n\frac{(E_0 - E_n)|\vec{p}_{\mathrm{on}}|^2}{E_0 - E_n - \omega_k} \tag{17.9}$$

is the level shift of the bound electron and

$$\Delta E_F = -\frac{e^2}{6\pi^2 m^2}\int_0^\infty d\omega_k\sum_n|\vec{p}_{\mathrm{on}}|^2 \tag{17.10}$$

is the energy shift of the free electron. With this separation, the divergence in ΔE_B is logarithmic. We can write the ΔE_F part as

$$\Delta E_F = -\frac{e^2 p^2}{6\pi^2 m^2}\int_0^\infty d\omega_k, \tag{17.11}$$

which corresponds to a mass correction δm of the free electron so that

$$E = \frac{p^2}{2(m + \delta m)} = \frac{p^2}{2m} + \Delta E_F. \tag{17.12}$$

From this identity we get

$$\delta m = -\frac{\Delta E_F}{p^2/2m + \Delta E_F} = \frac{e^2/3\pi^2 m \int_0^\infty d\omega_k}{1 - \frac{2e^2}{3\pi^2 m}\int_0^\infty d\omega_k}. \tag{17.13}$$

Actually, the $d\omega_k$ integrals in (17.9) and (17.11) should be cut off at an appropriate value since we are in the non-relativistic domain. In that case we get

$$\Delta E_B = \frac{e^2}{6\pi^2 m^2}\sum_n (E_n - E_0)|\vec{p}_{\text{on}}|^2 \log\left(\frac{\omega_k^{\max}}{E_n - E_0}\right). \tag{17.14}$$

If we now define $(E - E_0)_{\text{av}}$ by the identity

$$\sum_n (E_n - E_0)|\vec{p}_{\text{on}}|^2 \log(E_n - E_0) = \log(E - E_0)_{\text{av}} \sum_n (E_n - E_0)|p_{\text{on}}|^2, \tag{17.15}$$

then

$$\Delta E_B = \frac{e^2}{6\pi^2 m^2}\log\frac{\omega_k^{\max}}{(E - E_0)_{\text{av}}}\sum_n (E_n - E_0)|p_{\text{on}}|^2. \tag{17.16}$$

By making use of appropriate commutation relations, it is possible to get

$$\sum_n (E_n - E_0)|\vec{p}_{\text{on}}|^2 = \frac{e^2}{2}\left\langle 0 \left| \nabla^2 \frac{1}{r} \right| 0 \right\rangle = \frac{e^2}{2}\delta(\vec{r})|\psi_0|^2 \tag{17.17}$$

so that we have

$$\Delta E_B = \frac{e^4}{2\pi^2 m^2}|\psi_0(0)|^2 \log\frac{\omega_k^{\max}}{(E - E_0)_{\text{av}}}. \tag{17.18}$$

Since $\psi(0) \neq 0$ only for the S states, only these will be shifted. For the state $|0\rangle = |nS\rangle$,

$$|\psi_n(0)|^2 = \frac{1}{\pi(na_0)^3}. \tag{17.19}$$

In terms of K_n defined by

$$(E - E_n)_{\text{av}} = K_n, \tag{17.20}$$

the relation (17.18) then becomes

$$\Delta E_B(ns) = \frac{4\alpha^2}{3\pi m^2}\frac{1}{(na_0)^3}\log\frac{\omega_k^{\max}}{K_n}. \tag{17.21}$$

This result was derived by H.A. Bethe. If we take $\omega_k^{\max} = m$ and K_n calculated by Bethe, one gets

$$\Delta E_B(2S) = 1040\,\text{Mcps}. \tag{17.22}$$

The experimental value is

$$\Delta E_B(2S) = 1057.77 \pm 0.10\,\text{Mcps}. \tag{17.23}$$

17.3 Welton Method

As mentioned in the introduction, Welton considered the interaction of the electron with fluctuating vacuum of the quantized electromagnetic field which gives rise to fluctuating motion of the former whose equation of motion can be written as

$$m\delta\ddot{\vec{r}} = e\vec{\mathcal{E}}_{\text{vac}} = \sum_{\vec{k}\lambda} e\vec{\mathcal{E}}_{\vec{k}\lambda}\cos(\omega_{k\lambda}t - \vec{k}\cdot\vec{r}), \quad \lambda = 1, 2, \tag{17.24}$$

whose solution for $\vec{k}\cdot\vec{r} \ll 1$ is

$$\delta\vec{r} = -\frac{e}{m}\sum_{\vec{k}\lambda}\vec{\mathcal{E}}_{\vec{k},\lambda}\frac{\cos\omega_k t}{\omega_{k\lambda}} \tag{17.25}$$

from which we have

$$\overline{(\delta\vec{r})^2} = \frac{e^2}{2m^2}\sum_{\vec{k}\lambda}\frac{(\vec{\mathcal{E}}_{\vec{k},\lambda})^2}{(\omega_{k\lambda})^4} \tag{17.26}$$

since $\overline{\cos\omega t} = 0$, $\overline{\cos\omega t\cos\omega' t} = \frac{1}{2}\delta_{\omega\omega'}$.

The zero point energy of the field is

$$E = \frac{1}{2}\int d^3r\vec{\mathcal{E}}^2_{vac} = \frac{1}{2}\sum_{\vec{k},\lambda}\hbar\omega_{\vec{k}\lambda}, \tag{17.27}$$

which on using (17.24) gives

$$\vec{\mathcal{E}}^2_{\vec{k}\lambda} = \hbar\omega_{k\lambda}. \tag{17.28}$$

Substituting this in (17.26) gives

$$(\delta\vec{r})^2 = \frac{e^2\hbar}{2m^2}\sum_{\vec{k}\lambda}\frac{1}{\omega^3_{k\lambda}} = \frac{1}{2}\frac{e^2}{\hbar c}\left(\frac{\hbar}{mc}\right)^2\int_{\omega_{\min}}^{\omega_{\max}}\frac{d\omega}{\omega}. \tag{17.29}$$

By taking $\omega_{\max} = \dfrac{mc^2}{\hbar}$, $\omega_{\min} = \dfrac{me^4}{2n^2\hbar^3}$ and performing the integration, we get

$$\overline{(\delta\vec{r}^2)} = \frac{1}{2}\alpha\left(\frac{\hbar}{mc}\right)^2\ln\frac{2n^2}{\alpha^2}, \quad \alpha = \frac{e^2}{\hbar c} \tag{17.30}$$

$$r_{\text{vac}} = \left(\overline{\delta\vec{r}^2}\right)^{1/2} \sim \sqrt{\alpha}\frac{\hbar}{mc},$$

which means that the point electron acquires a finite size due to its interaction with vacuum fluctuations of the quantized electromagnetic field as a result of which the excess potential energy is

$$\begin{aligned}\delta V_{\text{vac}} &= -\,e\Phi(\vec{r}+\delta\vec{r}) + e\Phi(\vec{r})\\ &= -\,e\delta\vec{r}\cdot\vec{\nabla}\Phi - \frac{e}{2}(\delta\vec{r}\cdot\vec{\nabla}\Phi)^2 + \cdots .\end{aligned} \tag{17.31}$$

Since $\overline{\delta\vec{r}} = 0$ and

$$\overline{(\delta\vec{r}\cdot\vec{\nabla})^2} = \frac{1}{3}(\delta\vec{r})^2\nabla^2 = \frac{1}{3}r_{\text{vac}}\nabla^2,$$

$$\delta V_{\text{vac}} = \frac{e^2}{6}r_{\text{vac}}\nabla^2\Phi = \frac{4\alpha}{3}e^2\left(\frac{\hbar}{mc}\right)^2 \ln\frac{2n^2}{\alpha^2}\delta(\vec{r}),$$

$$\delta E_{\text{vac}} = \int d^3r\psi^*(\vec{r}\,')\delta V_{\text{vac}}\psi(\vec{r}\,') = \frac{4\alpha}{3}e^2\left(\frac{\hbar}{mc}\right)^2 |\psi(0)|\ln\frac{2n^2}{\alpha^2}. \tag{17.32}$$

For the state nS,

$$|\psi(0)|^2 = \frac{1}{\pi n^3 a_0^3} \tag{17.33}$$

so that

$$\delta E_{\text{vac}}(ns) = \frac{8\alpha^3}{3\pi n^3}R\ln\frac{2n^2}{\alpha^2}, \tag{17.34}$$

where $R = \dfrac{me^4}{2\hbar^2}$ is the Rydberg constant. For the state $2S$, this formula gives

$$\delta E_{\text{vac}}(2S) = 1040\,\text{Mcps}. \tag{17.35}$$

Problems

1. Calculate the Lamb shift of the ground state of hydrogen atom and compare it with that of the $2S$ state.

2. Obtain the Stark splitting of $n = 2$ state of hydrogen atom taking into account the Lamb shift.

   Hint: Take $H_I = \begin{pmatrix}\Delta & e\mathcal{E}a\\ e\mathcal{E}a & 0\end{pmatrix}$
   in the space of states $|200\rangle$ and $|210\rangle$. Here Δ = Lamb shift of $|200\rangle$ and $a = \langle 200|z|210\rangle$.
   Note that Lamb shift of $|210\rangle$ is zero. Obtain the eigenvalues and eigenfunctions of the above H_I.

Bibliography

[1] W.E. Lamb and R.C. Retherford, *Phys. Rev.*, **79**, 549 (1951); **86**, 1014 (1951).

[2] W.E. Lamb, *Rep. Prog. Phys.*, **14**, 19 (1951).

[3] H.A. Bethe, *Phys. Rev.*, **72**, 339 (1947).

[4] *Encyclopeadia of Physics*, ed. S. Flugge, Vol. XXXV, Atoms I, *Quantum Mechanics of One and Two-Electron Systems*, by H.A. Bethe and E.E. Salpeter (1957).

[5] A.A. Sokolov, I.M. Ternov and V. Ch. Zhukovskii, *Quantum Mechanics*, Mir Publishers, Moscow (1984).

[6] T.A. Welton, *Phys. Rev.*, **74**, 1157 (1948).

18 Theory of Scattering

18.1 Introduction

So far, we have applied quantum mechanics mostly to bound states. As far as scattering is concerned, we have devoted one chapter to the scattering of photons by free and bound particles. In this chapter we shall be concerned with the general theory of scattering. In section 2 we define scattering amplitude and cross-section with reference to the actual experimental setup and in section 3 we relate these to the Schrödinger wavefunction by making use of the wave equation. In section 4 we present Heisenberg's S-matrix theory and in section 5 and 6 discuss its unitarity and reciprocity properties. In section 7 we develop the Lippmann–Schwinger formulation of scattering theory. Sections 8 and 9 are devoted to partial wave analysis of elastic and inelastic scattering. In section 10 we present the exact treatment of Coulomb scattering and in section 11 and 12 we discuss the analytic properties of scattering amplitude in the complex momentum and angular momentum plane.

18.2 Schematic Presentation of Scattering Experiment

In a typical scattering experiment, schematically shown in Fig. 18.1, a beam of incident particle strikes a target and gets scattered in all directions; this can be represented as a spherical wave. The number of particles scattered per unit time into the solid angle $d\Omega$ per unit flux density of incoming particles is measured to give the differential scattering cross-section whose integral over the solid angle

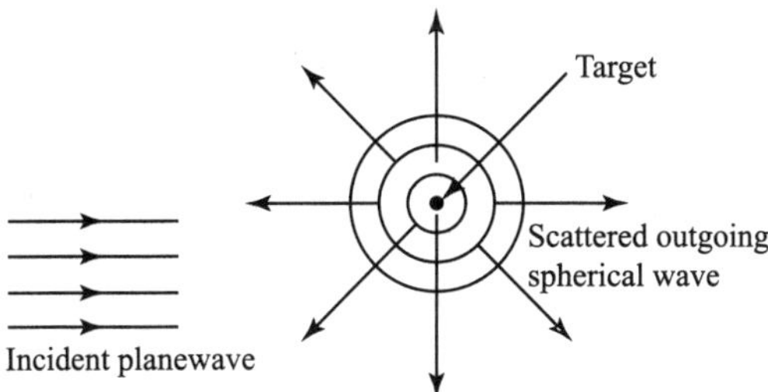

Fig. 18.1 *Schematic diagram of scattering experiment*

gives the total scattering cross-section. The distances of the sources of incident particles and the instruments recording the scattered particles from the target in the laboratory can be considered to be very large in the atomic scale; in fact they are taken to be infinite. One can therefore represent the wavefunction of the scattering system at large distances from the target as

$$\psi = \psi_{\text{in}} + \psi_{\text{sc}} = e^{ikz} + f(\theta)\frac{e^{ikr}}{r}, \tag{18.1}$$

where the first term represents the incident plane wave and the second term represents the scattered spherical wave whose amplitude $f(\theta)$ is called scattering amplitude. From (18.1) we have, for the number of scattered particles per unit time per unit solid angle,

$$n_{\text{sc}} = r^2 J_r = \frac{\hbar}{2im}\left(\psi_{\text{sc}}^* \frac{\partial \psi_{\text{sc}}}{\partial r} - \frac{\partial \psi_{\text{sc}}^*}{\partial r}\psi_{\text{sc}}\right) = \frac{\hbar k}{m}|f(\theta)|^2, \tag{18.2}$$

and the number of incident particles crossing unit area per unit time,

$$n_{\text{in}} = \frac{\hbar}{2im}\left(\psi_{\text{in}}^* \frac{\partial \psi_{\text{in}}}{\partial z} - \frac{\partial \psi_{\text{in}}^*}{\partial z}\psi_{\text{in}}\right) = \frac{\hbar k}{m}. \tag{18.3}$$

The differential cross-section, $\frac{d\sigma}{d\Omega}$, which is the ratio between the two, works out to

$$\frac{d\sigma}{d\Omega} = \frac{n_{\text{sc}}}{n_{\text{in}}} = |f(\theta|^2. \tag{18.4}$$

The total cross-section σ is obtained by integrating over the solid angle:

$$\sigma = \int \frac{d\sigma}{d\Omega} d\Omega = \int d\Omega |f(\theta)|^2. \tag{18.5}$$

18.3 Scattering Amplitude from Wave Equation

The Schrödinger equation for scattering of particles by a target potential field $V(r)$ is

$$(\nabla^2 + k^2)\psi = \frac{2mV}{\hbar^2}\psi, \qquad k^2 = \frac{2mE}{\hbar^2}, \tag{18.6}$$

whose solution at a large distance from the target should be

$$\lim_{r\to\infty} \psi(r) = \frac{e^{i\vec{k}\cdot\vec{r}}}{(2\pi)^{3/2}} + \frac{e^{ikr}}{r} f(\theta). \tag{18.7}$$

At arbitrary distance the formal solution of (18.6) can be written as

$$\psi(\vec{r}) = \phi(\vec{r}) + \frac{2m}{\hbar^2} \int d^3r' \mathcal{G}_k^{(+)}(\vec{r} - \vec{r}') V(r') \psi(\vec{r}'), \tag{18.8}$$

where the Green's function $\mathcal{G}_k^{(+)}(\vec{r}')$ satisfies the equation

$$(\nabla^2 + k^2)\mathcal{G}_k^{(+)}(\vec{r}') = \delta(\vec{r}). \tag{18.9}$$

Solution of (18.9) gives

$$\mathcal{G}_k^{(+)}(\vec{r}) = -\frac{1}{4\pi} \frac{e^{ik|\vec{r}|}}{|\vec{r}|}, \tag{18.10}$$

which is an outgoing spherical wave. Substitution of (18.10) in (18.8) gives

$$\psi(\vec{r}) = \phi(\vec{r}) - \frac{m}{2\pi\hbar^2} \int d^3r' \frac{e^{ik|\vec{r}-\vec{r}'|}}{|\vec{r} - \vec{r}'|} V(\vec{r}') \psi(\vec{r}'), \tag{18.11}$$

which in the limit of large r becomes

$$\lim_{r\to\infty} \psi(\vec{r}) = \phi(\vec{r}) - \frac{m}{2m\hbar} \frac{e^{ikr}}{r} \int d^3r' e^{-i\vec{k}_f\cdot\vec{r}} V(\vec{r}') \psi(\vec{r}') \tag{18.12}$$

since in this limit

$$k|\vec{r} - \vec{r}'| = kr - k\frac{\vec{r}\cdot\vec{r}'}{r} = kr - \vec{k}_f \vec{r}'$$

with $\vec{k}_f = k\vec{r}/r$. Comparing (18.12) with (18.7) gives

$$f(\theta) = -\frac{m}{2\pi\hbar^2} \int d^3r e^{-i\vec{k}_f\cdot\vec{r}} V(\vec{r}) \psi_i(\vec{r}). \tag{18.13}$$

In order to evaluate $f(\theta)$, one needs to know ψ, which is determined by the integral equation (18.11). The simplest method of obtaining the solution of this equation is that of iteration, which consists of the substitution of $\psi(\vec{r}')$ inside the integral of this equation by its value given by the two terms of the right-hand side of the equation itself and then go on repeating the process. This gives a series.

$$\psi(\vec{r}) = \phi(\vec{r}) - \frac{m}{2\pi\hbar^2} \int d^3r' \frac{e^{ik|\vec{r}-\vec{r}'|}}{|\vec{r} - \vec{r}'|} V(r') \phi(\vec{r}') + \left(\frac{m}{2\pi\hbar^2}\right)^2 \times \int d^3r' \int d^3r'' \frac{e^{ik|\vec{r}-\vec{r}'|}}{|\vec{r} - \vec{r}'|} V(r') \frac{e^{ik|\vec{r}'-\vec{r}''|}}{|\vec{r}' - \vec{r}''|} V(r'') \phi(r'') + \cdots. \tag{18.14}$$

Substituting $\phi(r)$ for $\psi(\vec{r})$ in (18.13) gives

$$f_B(\theta = -\frac{m}{2\pi\hbar^2}\int d^3r'\, e^{-i\vec{k}_f\cdot\vec{r}'}V(\vec{r}')\phi_i(\vec{r}')$$

$$= -\frac{m}{2\pi\hbar^2}\int d^3r'\, e^{-i(\vec{k}_i\cdot\vec{r}'_f)\cdot\vec{r}'}V(\vec{r}')\phi_i(\vec{r}'), \qquad (18.15)$$

which amounts to what is known as first Born approximation. Higher Born approximations can be obtained by substituting the successive terms in the iterated solution (18.14). For the case of scattering by a Coulomb potential $V(r) = -e^2/r$, calculation gives $f_B(\theta) = \frac{me^2}{2\hbar^2k^2}\frac{1}{\sin^2\frac{\theta}{2}}$.

It is clear that the Born approximation is valid if

$$|\phi(\vec{r})| \gg \frac{m}{2\pi\hbar^2}\left|\int d^3r'\,\frac{e^{ik|\vec{r}-\vec{r}'}}{|\vec{r}-\vec{r}'|}V(r')\phi(r')\right|. \qquad (18.16)$$

Usually, $V(r)$ is maximum at $r = 0$, in which case (18.16) boils down to

$$\frac{m}{2\pi\hbar^2}\left|\int d^3r\,\frac{V(r)}{r}\,e^{ikr+i\vec{k}_i\cdot\vec{r}}\right| \ll 1, \qquad (18.17)$$

which is

$$\left|\int_0^\infty dr\,V(r)[e^{2ikr}-1]\right| \ll \frac{k\hbar^2}{m}. \qquad (18.18)$$

For a potential having a range a such that $ka \ll 1$, i.e., for low energy, (18.17) reduces to

$$\frac{1}{4\pi a^2}\left|\int d^3r\frac{V(r)}{r}\right| \ll .\frac{\hbar^2}{2ma^2}. \qquad (18.19)$$

Since kinetic energy $= \hbar^2/2ma^2$ as per uncertainty relation, this amounts to

kinetic energy $\gg$ potential.

When the two particles colliding with each other are identical, it is not possible to distinguish between the two situations depicted in the two figures given in Fig. 18.2.

While in Fig. 18.2(a) the particle A is scattered through an angle θ, in Fig. 18.2(b) it is scattered through an angle $\pi - \theta$. Since the particles are identical ($A = B$), the two figures describe the same scattering. Hence one has to add $f(\theta)$ and $f(\pi-\theta)$ and then take the square modulus of the sum to obtain the scattering cross-section:

$$\frac{\delta\sigma}{d\Omega} = |f(\theta) + f(\pi-\theta)|^2. \qquad (18.20)$$

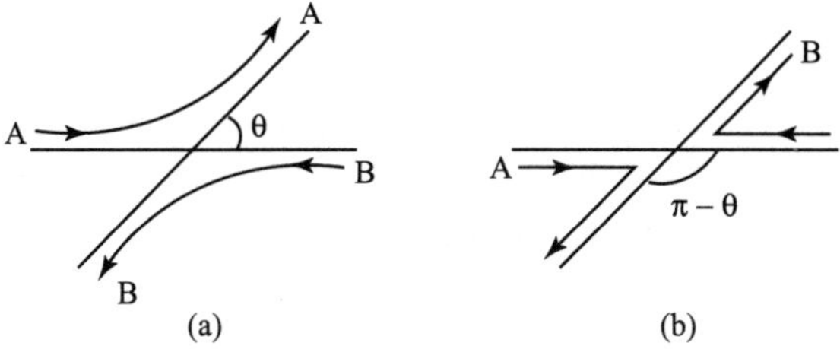

Fig. 18.2 *Collision between two identical particles: (a) direct (b) exchange*

For pure Coulomb scattering, this gives the Mott scattering formula.

$$\frac{d\sigma}{d\Omega} = \left(\frac{Z^2e^2}{2\mu v^2}\right)^2 \left[\frac{1}{\sin^4 \frac{1}{2}\theta} + \frac{1}{\cos^4 \frac{1}{2}\theta} + \frac{2\cos\left(\frac{Z^2e^2}{\hbar v}\ln\tan^2\frac{1}{2}\theta\right)}{\sin^2\frac{1}{2}\theta\cos^2\frac{1}{2}\theta}\right].$$

18.4 Heisenberg's S-Matrix

The wavefunction for elastic scattering can be presented in a manner somewhat different from section 1. Instead of the expression (18.1), we have

$$\psi(\vec{k},\vec{k}') = a(k)e^{ikr\lambda} + \frac{e^{ikr}}{r} f(\vec{k},\vec{k}')a(k), \tag{18.21}$$

where $|\vec{k}| = |\vec{k}'|$ and $\lambda = z/r$. Upon integrating over $d\Omega_k$ we get

$$\int \psi(\vec{k},\vec{k}')d\Omega_k = 2\pi \int_{-1}^{1} d\lambda a(\vec{k})e^{ikr\lambda},$$

$$+\frac{e^{ikr}}{r}\int d\Omega_k f(\vec{k},\vec{k}'), a(\vec{k}). \tag{18.22}$$

Since for $r \to \infty$ the function $e^{ikr\lambda}$ oscillates rapidly except at $\lambda = \pm 1$, the only nonvanishing contribution to the integral in the first term comes from near these two points and we have

$$\Psi(k) = \int d\Omega_k \psi(\vec{k},\vec{k}') = 2\pi\frac{i}{k}\left[\frac{e^{-ikr}}{r}a(-k) - \frac{e^{ikr}}{r}a(k)\right]$$

$$+\frac{e^{ikr}}{r}\int d\Omega_k f(\vec{k},\vec{k}')a(\vec{k}), \tag{18.23}$$

which can be written in the form

$$\int d\Omega_k \psi(\vec{k},\vec{k}') = \frac{i}{k}\left[\frac{e^{-ikr}}{r}a(-\vec{k}) - \frac{e^{ikr}}{r}Sa(\vec{k})\right], \tag{18.24}$$

with

$$Sa(\vec{k}) = a(\vec{k}) + \frac{ik}{2\pi} \int_{-1}^{+} d\Omega_k f(\vec{k}, \vec{k}').a(\vec{k}). \tag{18.25}$$

Here S is in the form of an integral operator,

$$S = 1 + 2ikf \quad \text{with} \quad fa = \frac{1}{4\pi} \int d\Omega_k f(\vec{k}, \vec{k}')a_k, \tag{18.26}$$

which is Heisenberg's scattering operator and is often referred to as the S-matrix. There are two terms on the right-hand side of (18.24), the first term representing an incoming spherical wave and the second term an outgoing spherical wave. This tells us that scattering can be viewed as conversion of an incoming spherical wave to an outgoing spherical wave, the latter having been affected by the interaction with the target. The extent to which it is affected is represented by the S-matrix.

18.5 Unitarity of S-Matrix

One of the physical requirements of the scattering process is that the incoming flux of the particle must equal the outgoing flux, i.e.,

$$r^2 J_r^{\text{in}} = r^2 J_r^{\text{out}}, \tag{18.27}$$

where J_r is defined in (18.2). Calculation using

$$\psi_{\text{in}} = \frac{e^{-ikr}}{r} a(-k), \quad \psi_{\text{out}} = \frac{e^{ikr}}{r} Sa(k) \tag{18.28}$$

in (18.27) gives

$$r^2 J_r^{\text{in}} = \frac{k\hbar}{2m} |a|^2 \, , \; r^2 J_r^{\text{out}} = \frac{k\hbar}{2m} S^+ S |a|^2, \tag{18.29}$$

and therefore,

$$S^+ S = 1 \tag{18.30}$$

which tells that the S matrix must be unitary. This is a consequence of conservation of flux; in other words, conservation of probability. Condition (18.30), on account of (18.26), amounts to

$$f - f^+ = 2ikf^+ f. \tag{18.31}$$

Written out explicity, this equation becomes

$$f(\vec{k}, \vec{k}') - f^*(\vec{k}', \vec{k}) = \frac{ik}{2\pi} \int d\Omega_{k''} f^*(\vec{k}, \vec{k}'') f(\vec{k}', \vec{k}''). \tag{18.32}$$

If in this equation we set $\vec{k} = \vec{k}'$ which corresponds to forward scattering, we get

$$\text{Im } f(\vec{k}, \vec{k}) = \frac{k}{4\pi} \int d\Omega_{k''} |f^*(\vec{k}, \vec{k}'')|^2 = \frac{k\sigma}{4\pi}. \tag{18.33}$$

This is the optical theorem which was first derived by Bohr, Pierls and Placzek. It is thus a consequence of the unitarity of S-matrix.

18.6 Reciprocity of S-Matrix

Reciprocity of the S-matrix expressed by the equation

$$S(\vec{k}, \vec{k}') = S(-\vec{k}', -\vec{k}) \tag{18.34}$$

follows from time reversal properties of the scattering states $|\vec{k}\rangle$ and $|\vec{k}'\rangle$. Writing

$$S(\vec{k}, \vec{k}') = \langle \Phi(\vec{k}') | S | \Phi(k) \rangle \tag{18.35}$$

and

$$S(-\vec{k}', -\vec{k}) =< \Phi(-\vec{k}) | S | \Phi(-\vec{k}'), \tag{18.36}$$

and noting that under time reversal,

$$T\Phi(\vec{k}) = \Phi(-\vec{k}) = \Phi^*(k), \tag{18.37}$$

where T is the time reversal operator, for $H = H^*$, which is true for spinless particles, we obtain

$$\begin{aligned} S(-\vec{k}', -\vec{k}) &= \langle \Phi^*(\vec{k}) | S | \Phi^*(\vec{k}') \rangle \\ &= \langle \Phi(\vec{k}') | S | \Phi^*(\vec{k}) \rangle = S(\vec{k}, \vec{k}'). \end{aligned} \tag{18.38}$$

According this 'reciprocity theorem', as it is called, scattering matrix and its time reversed matrix are the same. Time reversal reverses the direction of motion of the particles and interchanges the initial and the final state.

Since under space inversion or parity transformation

$$P|\Phi_k(\vec{k})\rangle = |\Phi(-\vec{k})\rangle \tag{18.39}$$

and if the Hamiltonian is invariant under this transformation, it is clear that under simultaneous operation of space and time inversion, $\Phi(\vec{k}) \to \Phi(\vec{k})$ and hence

$$S(\vec{k}'\vec{k}) = S(\vec{k}, \vec{k}'), \tag{18.40}$$

which is called detailed balancing property of the S-matrix.

18.7 Lippmann–Schwinger Formalism

An elegant theory of scattering was formulated by Lippmann and Schwinger which we present here. The Schrödinger equation for scattering of a particle by a potential V is

$$(E - H_0)\psi = V\psi. \tag{18.41}$$

In the absence of the potential, i.e., at large distance from the target, the equation is

$$(E - H_0)\phi = 0. \tag{18.42}$$

These two equations can be combined into a single equation which reads as

$$\psi = \phi + \frac{1}{E - H_0} V\psi. \tag{18.43}$$

The second term in this equation has a singularity which is defined as:

$$\psi^{(\pm)} = \phi + \frac{1}{E - H_0 \pm i\epsilon} V(\psi^{(\pm)}. \tag{18.44}$$

This equation is known as Lippmann–Schwinger equation, which for $\psi^{(+)}$ can be written as

$$\psi_{\vec{k}}^{(+)} = \phi_{\vec{k}} + \sum_{\vec{k}'} \frac{1}{E - H_0 + i\epsilon} |\phi_{\vec{k}'} > \langle \phi_{\vec{k}'}|V|\psi_{\vec{k}}^{+}\rangle \tag{18.45}$$

$$= \phi_{\vec{k}}(\vec{r}) + \int \frac{d^3\kappa'}{(2\pi)^3} \frac{e^{i\vec{k}'\cdot\vec{r}}}{E_{\vec{k}} - E_{\vec{k}'} + i\epsilon} \langle \vec{k}'|T|\vec{k}\rangle, \tag{18.46}$$

where

$$\langle \vec{k}'|T|\vec{k}\rangle = \langle \phi_{\vec{k}'}|V|\psi_{\vec{k}}^{(+)}\rangle. \tag{18.47}$$

On performing the angular integration in the second term in the manner it was done in section 2 for large r, (18.46) takes the form

$$\psi_{\vec{k}}^{(+)}(\vec{r}) = \phi_{\vec{k}}(\vec{r}) - i\pi \int_{-\infty}^{+\infty} \frac{k'dk'}{E_k - E_{k'} + i\epsilon}$$

$$\times \left[\frac{e^{ikr}}{r} \langle k\hat{r}|T|\vec{k}\rangle - \frac{e^{-ikr}}{r} \langle -k\hat{r}|T|\vec{k}\rangle \right]. \tag{18.48}$$

The k' integration can be performed by the standard method of choosing appropriate contour in the analytically-continued complex k' space. The position of the pole is such that only the first term contributes and we have

$$\psi_k^{(+)}(\vec{r}') = \phi_{\vec{k}'}(\vec{r}) + 4\pi^2 k \frac{dk}{dE_k} \langle k\hat{r}|T|\vec{k}\rangle \frac{e^{ikr}}{r}. \tag{18.49}$$

Comparison of this result with (18.1) gives

$$f(\theta) = 4\pi^2 k \frac{dk}{dE} \langle \phi_{\vec{k}'}|V|\psi_{\vec{k}}^{(+)}\rangle. \tag{18.50}$$

For the incoming state $\psi_{\vec{k}}^{(-)}(\vec{r})$, we put $(-i\epsilon)$ in the place of $(i\epsilon)$ in all the formulae above so that only the second term in (18.48) contributes and we have

$$\psi_{\vec{k}}^{(-)}(r) = \phi_{\vec{k}}(\vec{r}') + 4\pi^2 k \frac{dk}{dE} \langle -k\hat{r}|T|\vec{k}\rangle \frac{e^{-ikr}}{r}. \tag{18.51}$$

It will be seen from (18.49) and (18.51) that while $\psi_{\vec{k}}^{(+)}(\vec{r})$ is a sum of a plane wave $\phi_{\vec{k}}(\vec{r})$ and an outgoing spherical wave $\frac{e^{ikr}}{r}$, $\psi_{\vec{k}}^{(-)}(\vec{r})$ is a sum of plane wave $\phi_{\vec{k}}(\vec{r})$ and an incoming spherical wave $\frac{e^{-ikr}}{r}$. In the Lippmann–Schwinger formalism, the S-matrix turns out to be the scalar product of $\psi^{(+)}$ and $\psi^{(-)}$:

$$S_{fi} = \langle \psi_f^{(-)}|\psi_i^{(+)}\rangle. \tag{18.52}$$

In order to show this, we note that $\psi^{(\pm)}$ can also be written as

$$\psi^{(\pm)} = \phi + \frac{1}{E - H \pm i\epsilon} V\phi, \tag{18.53}$$

from which we get

$$-\psi_f^{(+)} + \psi_f^{(-)} = 2\pi i \delta(E_f - H) V\phi_f. \tag{18.54}$$

Taking scalar product of this with $\psi_i^{(+)}$, we get

$$-\langle \psi_f^{(+)}|\psi_i^{(+)}\rangle + \langle \psi_f^{(-)}|\psi_i^{(+)}\rangle = 2\pi i \delta(E_f - H)\langle \phi_f|V|\psi_i^{(+)}\rangle.$$

Noting that

$$\langle \psi_f^{(+)}|\psi_i^{(+)}\rangle = \delta_{if},$$

we have

$$\begin{aligned}\langle \psi_f^{(-)}|\psi_i^{(+)}\rangle &= \delta_{if} + 2\pi i \delta(E_f - H)\langle \phi_f|V|\psi_i^{(+)}\rangle \\ &= \delta_{fi} + \langle f|T|i\rangle,\end{aligned} \tag{18.55}$$

with

$$\begin{aligned}\langle f|T|i\rangle &= 2\pi \delta(E_f - E_i)\langle \phi_f|V|\psi_i^{(+)}\rangle \\ &= 2\pi i \delta(E_f - E_i)\langle f|t|i\rangle,\end{aligned} \tag{18.56}$$

which is equivalent to (18.26) if $\langle\psi_f^{(-)}|\psi_i^{(+)}\rangle$ is identified as the S-matrix. The differential cross-section is found from probability of transition P_{fi} from the initial to the final state, which is

$$
\begin{aligned}
P_{fi} &= |\langle f|T|i\rangle|^2 = 4\pi^2|t_{fi}|^2)\delta(E_i - E_f))^2 \\
&= 4\pi^2|t_{fi}|^2\delta(E_i - E_f)\lim_{\tau-\infty}\frac{1}{2\pi}\int_{-\tau/2}^{+\tau/2} dte^{\frac{i}{\hbar}(E_i-E_f)t} \\
&= 2\pi\tau|t_{fi}|^2\delta(E_i - E_f)\frac{\sin(E_i - E_f)\tau}{(E_i - E_f)\tau} \\
&= 2\pi\tau|t_{fi}|^2\delta(E_i - E_f).
\end{aligned}
\tag{18.57}
$$

This gives the cross-section which is the transition probability per unit time per unit incident flux,

$$
\sigma_{fi} = \frac{2\pi m}{\hbar k}|t_{fi}|^2\delta(E_i - E_f). \tag{18.58}
$$

18.8 Method of Partial Waves for Elastic Scattering

The scattering amplitude can be expanded in terms of the complete set of Legendre polynomials:

$$
f(\theta) = \sum_l (2l+1) f_l P_l(\cos\theta). \tag{18.59}
$$

The coefficient f_l is called partial wave scattering amplitude. This can be related to the partial wave S-matrix S_l which gives us the extent to which the outgoing spherical partial wave component of the incident plane wave is affected by the potential. To see this, we also expand the scattering wave function ψ in Legendre polynomials:

$$
\psi(r) = \sum_l A_l R_{kl}(r) P_l(\cos\theta) \tag{18.60}
$$

and substitute it in the Schrödinger equation

$$
(\nabla^2 + k^2)\psi = \frac{2mV}{\hbar^2}\psi \tag{18.61}
$$

to obtain the radial equation

$$
\left[\frac{d^2}{dr^2} + k^2 - \frac{l(l+1)}{r^2}\right] R_{kl}(r) = \frac{2mV}{\hbar^2} R_{kl}(r). \tag{18.62}
$$

Its asymptotic solution

$$R_{kl}(r) \xrightarrow[r\to\infty]{} \frac{2}{r}\sin\left(kr - \frac{l\pi}{2} + \delta_l\right\}$$

$$= \frac{1}{ir}\left\{(-i)^l e^{i(kr+\delta_l)} - (i)^l e^{-i(kr+\delta_l)}\right\}, \tag{18.63}$$

on substitution in (18.60) gives

$$\psi(r) \xrightarrow[r\to\infty]{} \sum_l -iA_l P_l(\cos\theta)\left\{(-i)^l \frac{e^{i(kr+\delta_l)}}{r} - \frac{(i)^l}{r} e^{-i(kr+\delta_l)}\right\}. \tag{18.64}$$

Since we want this to be of the form

$$\psi(r) \xrightarrow[r\to\infty]{} e^{ikz} + \frac{e^{ikr}}{r}, f(\theta) \tag{18.65}$$

and

$$e^{ikz} \xrightarrow[r\to\infty]{} \frac{1}{2ik}\sum_l (2l+1)P_l(\cos\theta)\left\{(-i)^{l+1}\frac{e^{-ikr}}{r} + \frac{e^{ikr}}{r}\right\}, \tag{18.66}$$

we must have

$$A_l = \frac{1}{2k}(2l+1)(i)^l e^{i\delta_l}, \tag{18.67}$$

in which case (18.64) takes the form

$$\psi(\vec{r}) \xrightarrow[r\to\infty]{} \frac{1}{2ki}\sum (2l+1)P_l(\cos\theta)\left\{(-i)^{l+1}\frac{e^{-ikr}}{r} + e^{2i\delta_l}\frac{e^{ikr}}{r}\right\} \tag{18.68}$$

$$= e^{ikz} + \sum_l (2l+1)P_l(\cos\theta)(e^{2i\delta_l} - 1)\frac{e^{ikr}}{r}, \tag{18.69}$$

so that

$$f(\theta) = \frac{1}{2ik}\sum_l (2l+1)P_l(\cos\theta)(e^{2i\delta_l} - 1). \tag{18.70}$$

It is seen from (18.66) and (18.68) that the partial outgoing spherical wave part of the plane wave is affected by a factor of

$$S_l = e^{2i\delta_l}, \tag{18.71}$$

which is called the partial wave scattering matrix. This is equivalent to a phase shift. The total scattering cross-section works out to be

$$\sigma = \int d\Omega \frac{d\sigma}{d\Omega} = \int d\Omega |f(\theta)|^2 = \sum_l \frac{4\pi}{k^2}(2l+1)\sin^2\delta_l. \tag{18.72}$$

From (18.70) one obtains

$$f(0) = \frac{1}{2ik}\sum_l (2l+1)(e^{2i\delta_l} - 1). \tag{18.73}$$

The optical theorem (18.33) then follows from (18.72) and (18.73).

Writing

$$\sigma = \sum_l \sigma_l \ , \quad f(\theta) = \sum_l (2l+1) P_l(\cos\theta) f_l, \tag{18.74}$$

we get

$$\sigma_l = \frac{4\pi}{k^2}(2l+1)\sin^2\delta_l, \tag{18.75}$$

$$f_l = \frac{1}{2ik}(e^{2i\delta_l} - 1). \tag{18.76}$$

The unitarity condition for f_l, that follows from this, is

$$\text{Im } f_l = k|f_l|^2, \tag{18.77}$$

which can be written as

$$\text{Im}\left(\frac{1}{f_l}\right) = -k. \tag{18.78}$$

In that case, we can put f_l in the form

$$f_l = \frac{1}{k(\cot\delta - i)}, \tag{18.79}$$

which is found useful in many applications. For low-energy scattering,

$$k\cot\delta = -\frac{1}{a} + \frac{1}{2}r_0 k^2 + \cdots, \tag{18.80}$$

where a is called the scattering length and r_0 the effective range.

From (18.75) we get

$$\sigma_l^{\max} = \frac{4\pi}{k^2}(2l+1), \tag{18.81}$$

which is larger than the classical value

$$\sigma_{l\ (\text{classical})} = \frac{\pi l^2}{k^2}. \tag{18.82}$$

This is due to the fact that in quantum mechanics particles are regarded as waves which exhibit interference.

The phase shift, which has been introduced as a parameter, can be determined by making use of the expansions

$$e^{-i\vec{k}_f\cdot\vec{r}} = \sum_l (2l+1)i^l R_{kl}(r)P_l(\cos\theta_f), \tag{18.83}$$

$$\psi_i = \frac{1}{kr}\sum_l (2l+1)i^l R_{kl}(r)P_l(\cos\theta_i), \tag{18.84}$$

where $\cos\theta_i = \frac{\vec{k}_i\cdot\hat{r}}{|\vec{k}_i|}$ and $\cos\theta_f = \frac{\vec{k}_f\cdot\hat{r}}{|\vec{k}_f|}$ of equation (18.13). The final result after some steps is

$$f(\theta) = -\frac{2m}{\hbar^2}\sum_l (2l+1)P_l(\cos\theta)\int_0^\infty drr^2V(r)R_{kl}(r)j_l(kr). \tag{18.85}$$

Comparing this with (18.13) we get

$$e^{i\delta_l}\sin\delta_l = \frac{-2mk}{\hbar^2}\int drr^2V(r)\ R_{kl}(r)j_l(kr). \tag{18.86}$$

This expression is exact; one needs to know $R_{kl}(r)$ for which the radial Schrödinger equation has to be solved. Born approximation discussed for the scattering amplitude amounts, in this case, to taking

$$R_{kl} \cong j_l(kr), \tag{18.87}$$

$$e^{i\delta_l}\sin\delta_l \cong \delta_l,$$

in which case, we get from (18.6)

$$\delta_l^{B\cdot A} = -\frac{2mk}{\hbar^2}\int_0^\infty drr^2V(r)j_l^2(kr). \tag{18.88}$$

18.9 Method of Partial Waves for Inelastic Scattering

If the target has internal structure which changes during collision so as to lose or gain energy, or new particles are produced as a result of collision, then

$$|S_l(k)| \neq 1,$$

which means, that instead of $S_l = e^{2i\delta_l}$ we need to have the representation

$$S_l(k) = \eta_l(k)e^{2i\delta_l(k)}, \qquad 0 \le \eta_l \le 1. \tag{18.89}$$

In that case,

$$f(\theta) = \frac{1}{2ik}\sum_l (2l+1)(\eta_l e^{2i\delta_l} - 1)P_l(\cos\theta), \tag{18.90}$$

and the elastic cross-section comes out to be

$$\begin{aligned}\sigma_{\text{el}} &= \frac{\pi}{2ik}\sum_l (2l+1)|(1-S_l)|^2 \\ &= \frac{\pi}{k^2}\sum_l (2l+1)(1+\eta_l^2 - 2\eta_l\cos 2\delta_l),\end{aligned} \tag{18.91}$$

which reduces to (18.72) when $\eta_l = 1$. To find the inelastic cross-section we start with (18.68):

$$\psi(\vec{r}) \xrightarrow[r\to\infty]{} \frac{1}{2ik}\sum_l (2l+1)P_l(\cos\theta)\left[(-1)^{l+1}\frac{e^{-ikr}}{r} + S_l\frac{e^{ikr}}{r}\right], \tag{18.92}$$

with S_l given by (18.89). From this we get

$$\text{Outgoing flux} = \frac{\pi}{k^2}\frac{\hbar k}{m}|S_l|^2, \tag{18.93}$$

$$\text{Incoming flux} = \frac{\pi}{k^2}\frac{\hbar k}{m},$$

so that

$$\text{the net loss of flux} = \frac{\hbar k}{m}\frac{\pi}{k^2}.(1-|S_l|^2).$$

Since the incident flux is hk/m, we have

$$\sigma_{\text{inel}} = \frac{\pi}{k^2}\sum_l (2l+1)(1-|S_l|^2) = \frac{\pi}{k^2}\sum_l (2l+1)(1-\eta_l^2). \tag{18.94}$$

Adding (18.91) and (18.94) gives

$$\sigma_{\text{tot}} = \sigma_{\text{el}} + \sigma_{\text{inel}} = \frac{2\pi}{k^2}\sum_l (2l+1)(1-\eta_l\cos 2\delta_l). \tag{18.95}$$

18.10 Coulomb Scattering

The exact solution of the Schrödinger equation for scattering of a charged particle by a repulsive Coulomb field can be obtained in parabolic coordinates. The scattering geometry being axially symmetric, the wavefunction is independent of the angle ϕ. Hence,

$$\psi(\xi,\eta) = f_1(\xi) f_2(\eta). \tag{18.96}$$

For $E > 0$, the Schrödinger equations for f_1 and f_2 then have the form

$$\frac{d}{d\xi}\left(\xi\frac{df_1}{d\xi}\right) + \left(\frac{1}{4}k^2\xi - \frac{1}{2} + \frac{\beta}{2}\right) f_1 = 0, \tag{18.97}$$

$$\frac{d}{d\eta}\left(\eta\frac{df_2}{d\eta}\right) + \left(\frac{1}{4}k^2\eta - \frac{1}{2} - \frac{\beta}{2}\right) f_2 = 0.$$

The incident plane wave is

$$\lim_{r\to\infty} \psi \sim e^{ikz} \qquad \text{for} \qquad -\infty < z < 0. \tag{18.98}$$

Noting that $r = \frac{\xi+\eta}{2}$ and $z = \frac{\xi-\eta}{2}$, condition (18.98) can be written in terms ξ and η as

$$\lim_{\eta\omega\infty} \psi(\xi\eta) = e^{\frac{ik}{2}(\xi-\eta)}. \tag{18.99}$$

Therefore we have

$$f_1(\xi) = e^{ik\xi/2}, \tag{18.100}$$

$$\lim_{\eta\to\infty} f_2(\eta) = e^{-ik\eta/2}.$$

Substituting $f_1(\xi)$ of (18.100) in the first of equation (18.97), we find that it is satisfied if

$$\beta - 1 - ik. \tag{18.101}$$

Substituting this value of β in the second of (8.97) we get

$$\frac{d}{d\eta}\left(\eta\frac{df_2}{d\eta}\right) + \left(\frac{1}{4}k^2\eta - 1 + \frac{ik}{2}\right) f_2 = 0. \tag{18.102}$$

Putting

$$f_2(\eta) = e^{-\frac{1}{2}ik\eta} W(\eta), \tag{18.103}$$

in (18.102) gives

$$\eta \frac{d^2W}{d\eta^2} + (1 - ik\eta)\frac{dW}{d\eta} - W = 0. \tag{18.104}$$

In term of a new variable $\lambda = ik\eta$, this equation reads

$$\lambda \frac{d^2W}{d\lambda^2} + (1 - \lambda)\frac{dW}{d\lambda} + \frac{W}{k} = 0, \tag{18.105}$$

which is a confluent hypergeometric equation. Its solution written in terms of η is

$$W(\eta) = CF\left(-\frac{i}{k}, 1, ik\eta\right). \tag{18.106}$$

Substitution of this in (18.103) to obtain $f_2(\eta)$ and the use of this and (18.100) in (18.96) gives

$$\psi(\xi, \eta) = Ne^{ik(\xi-\eta)}F\left(-\frac{i}{k}, 1, ik\eta\right). \tag{18.107}$$

The normalisation constant N is determined from the requirement that asymptotically the incident-wave part of it should be a plane wave with unit amplitude. In this limit,

$$e^{ik(\xi-\eta)}F\left(-\frac{i}{k}, 1, ik\eta\right) \to e^{\pi/2k}e^{ikz}\left(1 + \frac{1}{ik^3\eta}\right)e^{\frac{i}{k}\log k\eta}$$

$$- \frac{\frac{i}{k}e^{\pi/2k}e^{ik(\eta+z)}e^{-\frac{i}{k}\log k\eta}}{\Gamma\left(1 - \frac{i}{k}\right)ik\eta}. \tag{18.108}$$

It will be noticed that in this asymptotic form the incident-wave part is not just a plane wave e^{ikz} but a distorted plane wave $e^{ikz+\frac{i}{k}\log k\eta}$. The outgoing spherical wave part is not $\frac{e^{ikr}}{r}$, but once again distorted by a factor $e^{-\frac{i}{k}\log k\eta}$. This is due to the long range nature of the Coulomb potential. We can ignore these factors as their contribution to the incident and outgoing flux vanishes at infinity. Then we get from (18.108)

$$N = e^{-\pi/2k}\Gamma\left(1 + \frac{i}{k}\right), \tag{18.109}$$

$$f(\theta) - \frac{\Gamma\left(1 + \frac{i}{k}\right)}{\Gamma\left(1 - \frac{i}{k}\right)} \frac{e^{-2i/k \log \sin\theta/2}}{2k^2 \sin^2\theta/2}, \tag{18.110}$$

after making use of $\eta = r - z = r(1 - \cos\theta)$. The differential cross-section is

$$\frac{d\sigma}{d\Omega} = |f(\theta)|^2 = \frac{1}{4k^4 \sin^4 \theta/2}, \tag{18.111}$$

which is same as the one obtained from Born approximation as well as classical mechanics.

For an attractive Coulomb potential, the Schrödinger equations for f_1 and f_2 are

$$\frac{d}{d\xi}\left(\xi \frac{df_1}{d\xi}\right) + \left(\frac{1}{4}k^2\xi + \frac{1}{2} - \frac{\beta}{2}\right) f_1 = 0,$$

$$\frac{d}{d\eta}\left(\xi \frac{df_2}{d\eta}\right) + \left(\frac{1}{4}k^2\eta + \frac{1}{2} + \frac{\beta}{2}\right) f_2 = 0. \tag{18.112}$$

Substitution of (18.100) in these equations give

$$\beta = 1 + ik, \tag{18.113}$$

$$\frac{d}{d\eta}\left(\eta \frac{df_2}{d\eta}\right) + \left(\frac{1}{4}k^2\eta + 1 + \frac{ik}{2}\right) f_2 = 0. \tag{18.114}$$

Putting

$$f_2(\eta) = e^{-\frac{1}{2}ik\eta} W(\eta) \tag{18.115}$$

gives

$$\eta \frac{d^2W}{d\eta^2} + (1 - ik\eta)\frac{dW}{d\eta}{}_{W} = 0, \tag{18.116}$$

leading to

$$W(\eta) = CF\left(\frac{i}{k}, 1, ik\eta\right) \tag{18.117}$$

instead of (18.106) where the first argument of the hypergeometric function is $(-i/k)$. This finally leads to

$$\begin{aligned} f_{\text{attractive}}(\theta) &= \left(f_{\text{repuslive}}(\theta)\right)^* \\ &= -\frac{\Gamma(1 - \frac{i}{k})}{\Gamma(1 + \frac{i}{k})} \frac{e^{\frac{2i}{k}\log\sin\frac{\theta}{2}}}{2k^2 \sin^2\frac{\theta}{2}}. \end{aligned} \tag{18.118}$$

The partial wave amplitude can be found by substituting the phase shift δ_l obtained in equation (8.37) of chapter 8 in (18.70), where in the place of $(e^{2ik\delta_l} - 1)$

we put $e^{2i\delta_l}$ while only omitting unity since we do not consider forward scattering when the scattering amplitude diverges; for $\theta \neq 0$, this omitted term is

$$\frac{1}{2ik}\sum_l (2l+1)P_l(\cos\theta) = \frac{2}{ik}\delta(1-\cos\theta) = 0. \tag{18.119}$$

The final result is

$$f_l = e^{2i\delta_l} = \frac{\Gamma(l+1+\frac{i}{k})}{\Gamma(l+1-\frac{i}{k})}, \tag{18.120}$$

and for attractive potential

$$f_l = \frac{\Gamma(l+1-\frac{i}{k})}{\Gamma(l+1+\frac{i}{k})}. \tag{18.121}$$

18.11 Analytic Structure of S-Matrix in Complex Momentum Plane

The existence of the bound state, virtual (antibound) states, resonant (decaying) states and capture states of a particle with a target can be inferred by analytically continuing the S-matrix into the complex momentum plane.

The pole structure of the S-matrix in the complex momentum space is most elegantly analysed with the aid of Jost functions. In order to write S-matrix in terms of these functions, we note that two solutions, u_1 and u_2 of the radial Schrödinger equation for the s-wave ($l = 0$)

$$\frac{d^2u}{dr^2} + [k^2 - v(r)]u = 0 \tag{18.122}$$

where

$$u = \frac{R_0}{r} \quad \text{and} \quad v(r) = \frac{2mV(r)}{\hbar^2}, \tag{18.123}$$

are independent if their Wronskian

$$W(u_1, u_2) = u_1\frac{du_2}{dr} - \frac{du_1}{dr}u_2 \tag{18.124}$$

is non-zero. Two such solutions, $f(k,r)$ and $f(-k,r)$, with asymptotic form

$$\lim_{r\to\infty} f(kr) = e^{-ikr} \tag{18.125}$$

have their Wronskian

$$\lim_{r\to\infty} W(f(kr),\ f(-kr)) = 2ik. \tag{18.126}$$

The functions $f(k) = f(k,0)$ and $f(-k) = f(-k,0)$ are known as Jost functions. The scattering solutions of (18.122) can be written as

$$u_k(r) = -\frac{1}{2ik}\left[f(-k)f(kr) - f(k)f(-kr)\right]. \tag{18.127}$$

It follows from this and using (18.125) that asymptotically

$$u_k(r) \underset{r\to\infty}{\longrightarrow} \frac{-f(-k)}{2ik}\left[\frac{e^{-ikr}}{r} - \frac{f(k)}{f(-k)}\frac{e^{ikr}}{r}\right], \tag{18.128}$$

which means

$$R_k(r) \underset{r\to\infty}{\longrightarrow} \frac{-f(-k)}{2ik}\left[\frac{e^{-ikr}}{r} - \frac{f(k)}{f(-k)}\frac{e^{ikr}}{r}\right]. \tag{18.129}$$

It then follows that

$$S(k) = \frac{f(k)}{f(-k)} = e^{2i\delta(k)} \tag{18.130}$$

with the properies

$$S(-k) = S^{-1}(k),$$

$$\delta(-k) = -\delta(k). \tag{18.131}$$

It is possible to show that the zeroes of this S-matrix along the negative imaginary axis of the complex k-plane give us the energy of the bound states. Since these states have negative energy, we can set

$$k = -i\kappa.$$

For these imaginary momenta,

$$R_{\ i\kappa}(r) \underset{r\to\infty}{\longrightarrow} \frac{e^{-\kappa r}}{r} + S(-i\kappa)\frac{e^{\kappa r}}{r}. \tag{18.132}$$

The second term in the right-hand side of this equation blows up unless

$$S(-i\kappa) = 0, \tag{18.133}$$

in which case

$$R_{-i\kappa}(r) \underset{r\to\infty}{\longrightarrow} \frac{e^{-\kappa r}}{r} \tag{18.134}$$

and

$$E = \frac{k^2\hbar^2}{2m} = -\frac{\kappa^2\hbar^2}{2m}.$$

The asymptotic behaviour (18.134) is characteristic of bound states. Thus, we have condition (18.133) for the bound state which says that these states are obtained from the zero's of the S-matrix on the negative imaginary axis. As an example, we take the Coulomb scattering whose amplitude is

$$f(\theta) = -\frac{\Gamma(1-\frac{i}{k})}{\Gamma(1+\frac{i}{k})}\frac{e^{\frac{2i}{k}\log\sin\frac{\theta}{2}}}{2k^2\sin^2\frac{\theta}{2}} = \frac{\Gamma(\frac{-i}{k})}{\Gamma(\frac{i}{k})}\frac{e^{2ik}\log\sin\frac{\theta}{2}}{2k^2\sin^2\frac{\theta}{2}} \tag{18.135}$$

for an attractive potential. If we set $k = -i\kappa$, then $f(\theta)$ has zeroes corresponding to the poles of the gamma function at

$$\frac{1}{\kappa} = -n \qquad n = 1, 2, \cdots,$$

which correspond to

$$E = \frac{k^2}{2} = -\frac{\kappa^2}{2} = -\frac{1}{2n^2},$$

which are the bound-state energy eigenvalues obtained by the actual solution of the Schrödinger equation. However, all such zeroes may not always correspond to bound states. These are called redundant zeroes. Example of these poles occur for a particle scattered by a potential

$$V(r) = -\frac{m\alpha^2}{4\pi a^2}e^{-r/a}. \tag{18.136}$$

The radial wavefunction for the particle in this potential obtained by solution of the Schrödinger equation is, for positive energy,

$$R_k(r) = \frac{i}{\sqrt{2\pi}}\frac{\Gamma(1+i\rho)}{|J_{-i\rho}(\alpha)|}\left\{J_{-i\rho}(\alpha)J_{i\rho}(\alpha e^{-r/2a})J_{i\rho}(\alpha)J_{-i\rho}(\alpha e^{-r/2a})\right\} \tag{18.137}$$

with $\rho = 2ak$ whose asymptotic value is

$$R_k(r) \underset{r\to\infty}{=\!\!=\!\!\longrightarrow} \frac{i}{\sqrt{2\pi}}\frac{\Gamma(1+i\rho)}{|J_{-i\rho}(\alpha)|} \times\left\{\frac{J_{-i\rho}(\alpha)\left(\frac{\alpha}{2}\right)^{i\rho}}{\Gamma(1+i\rho)}e^{-kr} - \frac{J^{i\rho}(\alpha)\left(\frac{\alpha}{2}\right)^{-i\rho}}{\Gamma(1-i\rho)}e^{ikr}\right\}. \tag{18.138}$$

It then follows that

$$S(k) = \frac{J^{ip}(\alpha)\Gamma(1+i\rho)}{J_{-i\rho}(\alpha)\Gamma(1-i\rho)}\left(\frac{\alpha}{2}\right)^{-2i\rho}. \tag{18.139}$$

The bound states are obtained by demanding

$$S(-i\kappa) = 0,$$

which gives

$$\frac{J_{2a\kappa}(\alpha)\Gamma(1+2a\kappa)}{J_{-2a\kappa}(\alpha)\Gamma(1-wa\kappa)} = 0. \tag{18.140}$$

This is satisfied if either

$$J_{2a\kappa}(\alpha) = 0, \tag{18.141}$$

or

$$\Gamma(1-2a\kappa) = \infty. \tag{18.142}$$

This is what we get from the scattering solutions. On solving the Schrödinger equation for $E < 0$, one finds that only (18.141) corresponds to bound states, but not (18.142). The latter are called the redundant zeroes of the S-matrix.

One may be curious to know what the zeroes of the S-matrix on the positive imaginary axis correspond to. For this we put

$$S(i\kappa) = 0 \tag{18.143a}$$

in the asymptotic solution and find

$$R_{i\kappa} \xrightarrow[r\to\infty]{} \frac{e^{\kappa r}}{r} + S(i\kappa)\frac{e^{-\kappa r}}{r} = \frac{e^{\kappa r}}{r}. \tag{143b}$$

For $k = i\kappa$ as for $k = -i\kappa$,

$$E = -\frac{\kappa^2\hbar^2}{2m}. \tag{18.144}$$

Since the energy is negative, it is a bound state but the wavefunction blows up unless κ is infinitesimally small. This means that the zeroes on the positive imaginary axis of the complex momentum plane represent the bound state with energy almost equal to zero. Such states are virtual or anti-bound states. When such states exist, the S-matrix can be represented as

$$S(k) = \frac{\kappa + ik}{\kappa - ik} = e^{2i\delta}, \tag{18.145a}$$

which implies that

$$\kappa + ik = (k^2 + \kappa^2)^{1/2} e^{i\delta}, \tag{18.145b}$$

$$\sin\delta = \frac{k}{(k^2+\kappa^2)^{1/2}}, \tag{18.145c}$$

and therefore

$$\sigma = \frac{4\pi}{k^2} \sin^2 \delta = \frac{4\pi}{k^2 + k^2}. \tag{18.145d}$$

It may be noted here that for bound states which are poles on the positive imaginary k axis, we can represent the S-matrix by

$$S = \frac{\kappa - ik}{\kappa + ik}, \tag{18.146a}$$

$$\kappa - ik = (\kappa^2 + k^2)^{1/2} e^{-i\delta}, \tag{18.146b}$$

$$\sin \delta = \frac{k}{(k^2 + \kappa^2)^{1/2}}, \tag{18.146c}$$

$$\sigma = \frac{4\pi}{k^2 + \kappa^2}. \tag{18.146d}$$

Thus the total scattering cross-section for ($l = 0$) is the same for the bound state as for the virtual state. Therefore, if scattering cross-section for small k behaves like (18.145d) and there are no bound states with small binding energy, then it is an indication that virtual states exist.

The poles of the S-matrix in the fourth quadrant of the complex momentum plane correspond to decaying or resonant states. In this quadrant the momentum k is

$$k_r = k_r' - ik_r'' \quad k_r' > 0\,,\ k_r'' > 0, \tag{18.147}$$

and the energy is

$$E = \frac{k^2\hbar^2}{2m} = \left(\frac{k_r'^2}{2m} - \frac{k_r''^2}{2m} - \frac{ik_r'k_r''}{m} \right) \hbar^2$$

$$= E_r - \frac{i\Gamma_r}{2}, \tag{18.148}$$

where

$$E_r' = \frac{\hbar^2}{2m}(k_r'^2 - k_r''^2)\,,\quad \Gamma_r = \frac{2\hbar^2}{m} k_r' k_r''. \tag{18.149}$$

The Jost function for the lth partial wave can be represented by

$$f_l(k) = (k - k_r' - ik_r'')g_l,$$

$$f_l(-k) = (k - k_r' + ik_r'')g_l^*, \tag{18.150}$$

and the S-matrix by

$$S_l(k) = \frac{k - k'_r - ik''_r}{k - k'_r + ik''_r} e^{2i\delta_l}, \tag{18.151}$$

where

$$e^{2i\delta_l} = (-1)^l \frac{g_l}{g_l^*}. \tag{18.152}$$

In terms of energy,

$$S_l(E) = \frac{E - E_r - \frac{i\Gamma_r}{2}}{E - E_r + \frac{i\Gamma_r}{2}} e^{2i\delta_l} = \left[1 - \frac{i\Gamma_r}{E - E_r + \frac{i\Gamma_r}{2}}\right] e^{2i\delta_l}, \tag{18.153}$$

from which the cross-section for the lth partial wave scattering can be calculated and found to be

$$\sigma_l = \frac{\pi}{k^2}(2l+1)\left[4\sin^2\delta_l - 4\mathrm{Re}\left(e^{i\delta_l}\sin\delta_l \frac{\Gamma_r^2}{E - E_r + \frac{i\Gamma_r}{2}}\right)\right.$$
$$\left. + \frac{\Gamma_r^2}{(E - E_r)^2 + \frac{1}{4}\frac{\Gamma_r}{2}}\right]. \tag{18.154}$$

This is the Breit–Wigner formula. In this expression, the first term represents potential scattering and the third term resonance scattering; the second term is due to interference. It will be seen that

$$\sigma_l^{\max} = \frac{4\pi}{k^2}(l+1). \tag{18.155}$$

The result is graphically represented in the Fig. 18.3. Here the dotted line represents $\sigma = \pi/k^2$ and the solid line represents resonance scattering.

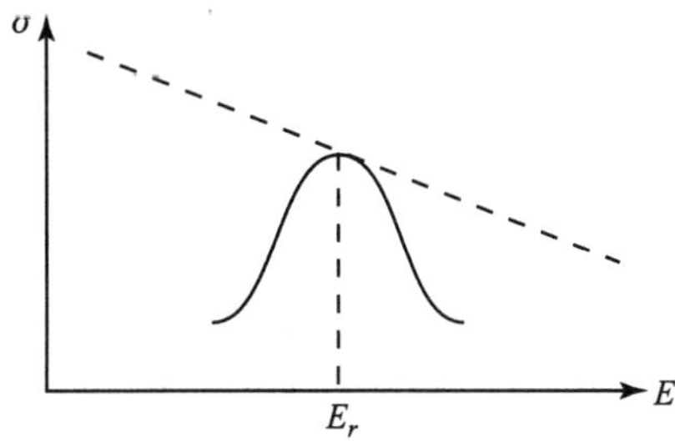

Fig. 18.3. *Resonance scattering (solid line) and potential scattering (dotted line) as function of energy E*

We can also calculate the phase shift and the scattering amplitude which are

$$\delta_l(E) = \delta_{\text{lr}} - \tan^{-1}\frac{\Gamma_r}{2(E - E_r)}, \tag{18.156}$$

$$f(\theta) = f_{\text{pot}} - \frac{2l+1}{2k}\frac{\Gamma_r}{E - E_r + \frac{i\Gamma_r}{2}}e^{2i\delta_{lr}}P_l(\cos\theta). \tag{18.157}$$

The poles in the third quadrant have momenta characterised by

$$k = -k_r' - ik_r'' \quad \text{with} \quad k_r' > 0 \quad , \quad k_r'' > 0. \tag{18.158}$$

The treatment here is thus the same as for the fourth quadrant with k_r' replaced by $-k_r'$, which amounts to replacing Γ_r by $-\Gamma_r$. Because of this, the time-dependence of the wavefunction will be

$$\psi \sim e^{-iE_rt + \frac{\Gamma_r}{2}t}, \tag{18.159}$$

instead of

$$\psi \sim e^{-iE_rt - \frac{\Gamma_r}{2}t} \tag{18.160}$$

for decaying or resonance states. The probability of the state

$$P \sim e^{\Gamma_r t} \tag{18.161}$$

at a given point will grow with time indicating thereby that matter is coming inside instead of getting lost as in the decaying state. In other words, capture is taking place. Therefore the poles in the third quadrant are called capture states.

The study of the poles of S-matrix in the complex momentum space can be summarised in Fig. 18.4.

The pole structure in the complex energy plane can be obtained from that of the complex momentum plane by noting that $E = k^2\hbar^2/2m$. We have, therefore, the correspondence given in Table 18.1. In accordance with this correspondence,

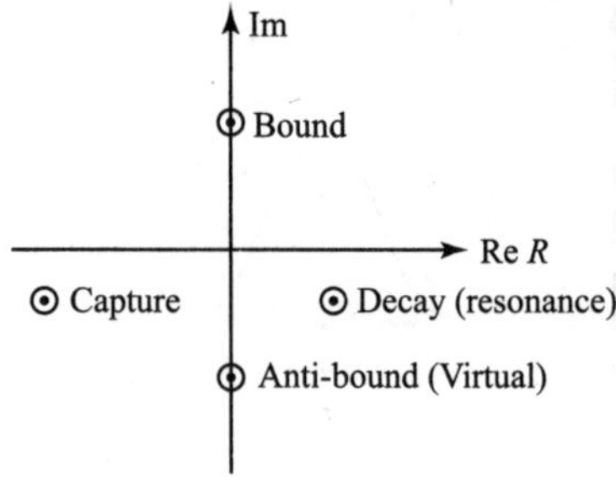

Fig. 18.4 *Poles of the S-matrix in the complex momentum plane*

Table 18. 1

	Momentum plane	Energy plane
1.	Positive imaginary axis	Negative real axis
2.	Negative imaginary axis	Negative real axis on the second Rieman sheet
3.	Third quadrant	Upper half plane of second Rieman sheet
4.	Fourth quadrant	Lower half plane of second Rieman sheet

the bound states will be poles on the negative real axis, the virtual states will be on the negative real axis in the second Rieman sheet, the capture states will be in the upper half of the second Rieman sheet and the resonance states in the lower half of the second Rieman sheet of the complex energy plane.

18.12 Analytic Structure of S-Matrix in Complex Angular Momentum Plane

In the previous section we have considered the analytic structure of the S-matrix continued to the complex momentum and energy planes. In this section we shall discuss its analytic structure in the complex l-plane. It is generally called the complex angular momentum plane. It should actually be called complex 'action plane' since l is the eigenvalue of the free particle 'action' rather than that of angular momentum as we have seen in chapter 9. The Schrödinger equation for this case takes the form

$$\frac{d^2u}{dr^2} + \left[\frac{2m}{\hbar^2}(E - V) - \frac{\alpha(\alpha+1)}{r^2}\right]u = 0, \tag{18.162}$$

where

$$u(r) = rR(r). \tag{18.163}$$

We first consider the negative energy solution so that $u(\infty) = 0$. Multiplying (18.162) by $u^*(r)$ and integrating over r by parts, we get, on using $u(\infty) = 0$,

$$-\int_0^\infty dr \left|\frac{du}{dr}\right|^2 + \frac{2m}{\hbar^2}\int_0^\infty dr(E - V)|u(r)|^2$$

$$-\alpha(\alpha+1)\int_0^\infty \frac{dr}{r^2}|u(r)|^2 = 0. \tag{18.164}$$

Since the first two terms are real, we have

$$\text{Im}\,\alpha(\alpha+1) = 0 \tag{18.165}$$

which gives

$$2\text{Re}\left(\alpha + \frac{1}{2}\right)\text{Im}\,\alpha = 0. \tag{18.166}$$

This tells us that

$$\text{Im}\,\alpha(\text{E}) = 0, \quad \text{for Re}\left(\alpha + \frac{1}{2}\right) > 0\ ,\ \ \text{E} < 0 \tag{18.167}$$

The condition $\text{Re}(\alpha + \frac{1}{2}) > 0$ is satisfied for $E < 0$ since for such states

$$R_\alpha(r) \underset{r\to 0}{\longrightarrow} c_1 r^\alpha + c_2 r^{-\alpha-1}. \tag{18.168}$$

To eliminate the second term, we require

$$\text{Re}\,\alpha > \text{Re}(-\alpha - 1) \tag{18.169}$$

which is satisfied if $\text{Re}(\alpha + \frac{1}{2}) > 0$.

In order to get conditions on $d\alpha/dE$, we differentiate the Schrödinger equation (18.162) with respect to energy and obtain

$$\frac{du''}{dE} + \frac{2m}{\hbar^2}(E - V)\frac{du}{dE} - \frac{2m}{\hbar^2}u$$
$$+ \frac{u}{r^2}\frac{d}{dE}[\alpha(\alpha+1)] - \frac{\alpha(\alpha+1)}{r^2}\frac{du}{dE} = 0. \tag{18.170}$$

Now we multiply (18.162) by $\frac{du}{dE}$ and (18.170) by u and then subtract. The result is

$$\frac{d}{dr}\left[\frac{du}{dr}\frac{du}{dE} - u\frac{d}{dr}\left(\frac{du}{dE}\right)\right] - \frac{2m}{\hbar^2}|u|^2 + \frac{|u|^2}{r^2}\frac{d}{dE}[\alpha(\alpha+1)] = 0. \tag{18.171}$$

We then integrate this equation over r by parts in the appropriate terms and obtain

$$\frac{d[\alpha(\alpha+1)]}{dE}\int_0^\infty dr\frac{|u(r)|^2}{r^2} = \frac{2m}{\hbar^2}\int_0^\infty dr|u(r)|^2. \tag{18.172}$$

Since $\text{Re}(\alpha + \frac{1}{2}) > 0$ for $E < 0$ that we are considering, we get from (18.172),

$$\frac{d[\alpha(\alpha+1)]}{dE} = 2\left(\alpha + \frac{1}{2}\right)\frac{d\alpha}{dE} > 0,$$

from which we obtain

$$\frac{d\alpha}{dE} > 0 \qquad \text{for } E < 0. \tag{18.173}$$

Next, we consider positive energy solutions, i.e. $E > 0$. For this we note that for complex l, the continuity equation becomes

$$\frac{\partial \rho}{\partial t} + \vec{\nabla} \cdot \vec{J} = 2\rho \text{Im } V_l > 0, \tag{18.174}$$

where

$$V_l = V + \frac{l(l+1)}{r^2} \tag{18.175}$$

is the effective potential. Equation (18.174) indicates that new particles are produced. For the dependence of α with energy, let us put

$$\alpha(E) = l_0 + i\eta + \beta(E - E_0), \tag{18.176}$$

where $\eta = \text{Im}\,\alpha(E_0), l_0 = \text{Re}\,\alpha(E_0),\ \beta = \left(\frac{d\alpha}{dE}\right)_{E=E_0}$. The asymptotic solutions for $u(r)$ can be written as (compare with (18.128)),

$$u_l = A(l, E)\frac{e^{-ikr}}{r} + B(l, E)\frac{e^{ikr}}{r}, \quad B(l^*, E) = A^*(l, E) \tag{18.177}$$

from which the partial wave S-matrix comes out as

$$S_l = \frac{B(l, E)}{A(l, E)}. \tag{18.178}$$

If we have a pole of S_l near $\alpha = l_0$, i.e.,

$$A(l_0, E) = \alpha(E) - l_0,$$

$$S_l = \frac{\beta(E - E_0) - i\eta}{\beta(E - E_0) + i\eta}. \tag{18.179}$$

On comparing with the Briet–Wigner formula (18.153),

$$S = \frac{E - E_0 - \frac{i\Gamma}{2}}{E - E_0 + \frac{i\Gamma}{2}} e^{2i\partial_e}, \tag{18.180}$$

we have,

$$\Gamma = \frac{2\eta}{\beta} = \frac{2\text{Im }\alpha}{\left(\frac{d\alpha}{dE}\right)_{E=E_0}}. \tag{18.181}$$

As an example of the theory presented above, we take the Coulomb scattering in an attractive potential for which (18.121), which is in Coulomb units, gives

$$S_\alpha = \frac{\Gamma\left(\alpha + 1 - \frac{ie^2}{\hbar}\sqrt{\frac{m}{2E}}\right)}{\Gamma\left(\alpha + 1 + \frac{ie^2}{\hbar}\sqrt{\frac{m}{2E}}\right)}. \tag{18.182}$$

The poles of this S-matrix are those of the gamma function in the numerator:

$$\alpha + 1 - \frac{ie^2}{\hbar}\sqrt{\frac{m}{2E}} = -n_r \ , \quad n_r = 0, 1, 2, \cdots . \tag{18.183}$$

This result was obtained by V. Singh. These poles are known as Regge poles. For $E > 0$, (18.183) gives

$$\text{Re}\,\alpha(E) = -n_r - 1,$$

$$\text{Im}\,\alpha(E) = \frac{e^2}{\hbar}\sqrt{\frac{m}{2E}}. \tag{18.184}$$

For $E < 0$, let $E = -\epsilon \ \epsilon > 0$.

$$\text{Re}\,\alpha(E) = -n_r - 1 + \frac{e^2}{\hbar}\sqrt{\frac{m}{2\epsilon}}, \tag{18.185}$$

$$\text{Im}\,\alpha(E) = 0.$$

If we put Re $\alpha = l$, (18.185) gives

$$E = -\frac{me^4}{2n^2\hbar^2}, \tag{18.186}$$

with $n = n_r + l + 1$ which is the Bohr's formula for energy levels of the hydrogen atom. The plot of Re $\alpha(E)$ versus E is known as the Regge trajectory. There are many such trajectories as shown in Fig. 18.5.

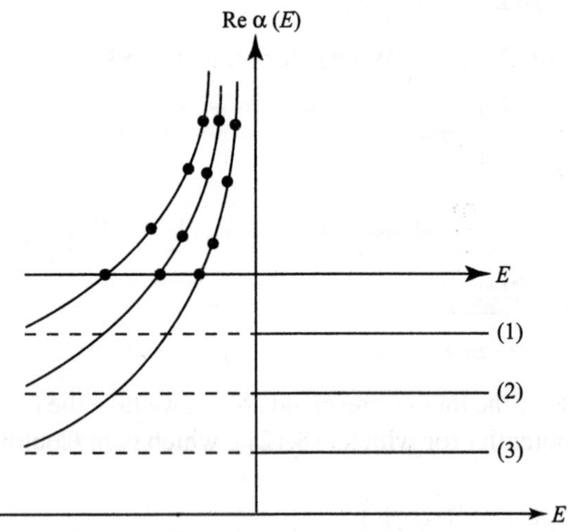

Fig. 18.5 *Regge trajectories for Coulomb scattering*

Problems

1. Obtain the optical theorem when inelastic processes are present.

2. Obtain δ_l for scattering of particles by a hard sphere. Show that for low-energy particles $\sigma = 4\pi a^2$, where a =radius of the sphere and further that this is four times the classical result. What is the reason for this enhancement?

3. Show by using partial wave expansion that the total cross-section for the scattering of particles by a back disc ($\eta_l = 0$) of radius a with sharp edges ($l \le ka$) is given by

$$\sigma = \sigma_{\text{el}} + \sigma_{\text{inel}} = 2\pi a^2.$$

4. By considering the contour integral

$$I = -\frac{1}{2\pi i} \cdot \frac{\oint df_l(k)/dk}{f_l(k)} dk$$

where the contour in the complex k-plane runs from $-\infty$ to $+\infty$ on the real axis and then returns back to $-\infty$ via a semicircle of infinite radius, show that

$$I = \frac{1}{\pi}\delta_l(k = 0) = n_l$$

where $f_l(k) = |f_l(k)|e^{i\delta_l(k)}$ and n_l is an integer.

5. Show that for complex k

$$f_l^*(-k^*) = f_l(k).$$

6. Obtain the condition of the validity of the Born approximation for the scattering of particles by a potential

$$V(r) = \begin{cases} -v & \text{for } r < a, \\ 0 & \text{for } r > a, \end{cases}$$

where v is independent of r.

Bibliography

[1] W. Heisenberg, *Zeits f. Physik*, **120**, 513, 673 (1943); *Zeits f. Natureforsch.* 11/12, 607 (1946).

[2] B.A. Lippmann and J. Schwinger, *Phys. Rev.*, **79**, 469 (1950).

[3] M.L. Goldberger and M. Gell-Man, *Phys. Rev.*, **91**, 398 (1953).

[4] R. Jost, *Helv. Physica Acta.*, **20**, 256 (1947).

[5] S.T. Ma, *Phys. Rev.*, **69**, 668 (1946); **71**, 195 (1947).

[6] C. Briet and E.P. Wigner, *Phys. Rev.*, **49**, 519 (1936).

[7] T. Regge, *Nuov Cimento.*, **14**, 451 (1959).

[8] V. Singh, *Phys. Rev.*, **127**, 632 (1967).

[9] L.D. Landau and E.M. Lifshitz, *Quantum Mehcanics*, Pergamon Press, Oxford (1977).

[10] N.F. Mott and H.S.W. Massey, *Theory of Atomic Collisions*, Oxford University Press (1949).

[11] A.C. Sitenko *Lectures in Scattering Theory*, Pergamon Press, Oxford (1971).

[12] N.F. Mott, *Proc. Roy. Soc.*, **A126**, 259 (1930).

[13] N. Bohr, R. Pierls and G. Placzek, *Nature*, **144**, 200 (1939).

[14] S.T. Ma, *Phys. Rev.*, **69**, 668 (1946); **71**, 159, 210 (1947)

[15] H.A. Bethe and R.F. Bacher, *Rev. Mod. Phys.*, **8**, 83 (1936)

19 Phase of the Wavefunction

19.1 Introduction

The state of a quantum mechanical system is described by a wavefunction which, in general, is complex and therefore can be represented by an amplitude and a phase

$$\psi(x) = \psi_0 e^{i\phi}. \tag{19.1}$$

Since the measurable quantities are integrals containing the wavefunction and its complex conjugate, they do not depend on the phase. The intensities in the interference experiments on the other hand are determined by the phase which, as we shall see, are integrals over potentials. We shall obtain the phases of particles with electric charge and magnetic dipole moments in scalar and vector potentials, and of neutral particles in earth's static gravitational field and the Coriolis force field due its rotation. We shall then discuss the problem in quantum mechanics in general as was done by Berry, for which reason this phase has been called the Berry phase. It is also called the geometrical or topological phase because of its dependence on the geometry of the system. We shall discuss the experimental measurement of the phase of the spin wave function of spin-$\frac{1}{2}$ particles and that of coherent states of macroscopic systems such as superconductors.

19.2 Charged Particle in a Farady Cage

Since electric field does not penetrate the inside of an electrically conducting sphere, a charged particle located in it does not experience any force. However, its wavefunction acquires a phase due to the effect of the electric potential:

$$i\hbar\frac{\partial\psi}{\partial t} = \left(\frac{p^2}{2m} + V(t)\right)\psi, \quad i\hbar\frac{\partial\psi_0}{\partial t} = \frac{p^2}{2m}\psi_0, \tag{19.2}$$

$$\psi(t) = \psi_0(t)\exp\left[\frac{-i}{\hbar}\int^t dt' V(t')\right] = \psi_0 e^{-iS/\hbar}. \tag{19.3}$$

As already mentioned, ordinarily this does not affect the measurable parameters since both $\psi(t)$ and its complex conjugate appear in the expectation values.

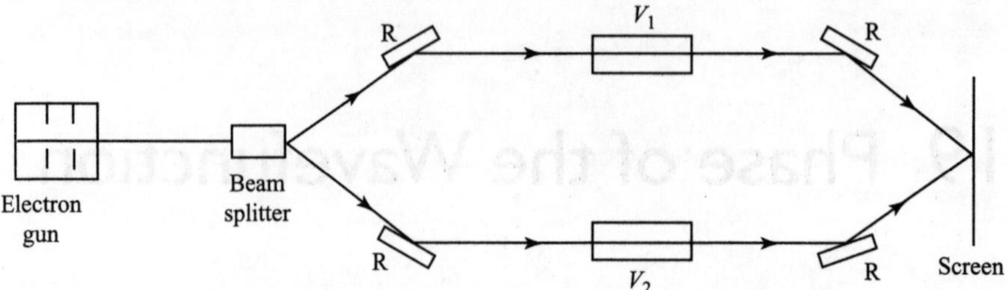

Fig. 19.1 *Interference between electron beams passing through Faraday cages*

However, the two charged particle beams emanating from a common source experiencing two different potentials in their paths as shown in the Fig. 19.1 exhibit interference effects when they are made to recombine.

$$\psi(t) = \psi_1^{(0)}(t)e^{-iS_1/\hbar} + \psi_2^{(0)}(t)e^{-iS_2/\hbar}, \tag{19.4}$$

$$S_i = \int dt V_i(t) \qquad i = 1, 2,$$

$$I = |\psi(t)|^2 = I_0\left[1 + \cos\left(\frac{S_1 - S_2}{\hbar}\right)\right], \tag{19.5}$$

since $\psi_1^{(0)} = \psi_2^{(0)}$, and

$$|\psi_1^{(0)}|^2 = |\psi_2^{(0)}|^2 = \frac{1}{2}\left(\psi_1^{*(0)}\psi_2^{(0)} + \psi_2^{(0)*}\psi_1^{(0)}\right) = \frac{I_0}{2}$$

as per geometrical arrangement. There will therefore be bright and dark spots (or lines) on the screen determined by

$$\Phi = \frac{S_1 - S_2}{\hbar} = 2n\pi \qquad \text{or} \qquad (2n+1)\pi. \tag{19.6}$$

19.3 Charged Particle in the Magnetic Field of a Solenoid

Bohm–Aharanov effect

Bohm and Aharanov considered the motion of a charged particle moving in the space outside of an infinitely long and thin cylindrical solenoid where the magnetic field is zero but the vector potential is non-zero, i.e., $\vec{A}$ is such that

$$\vec{B} = \vec{\nabla} \times \vec{A} = 0 \tag{19.7}$$

outside the solenoid. The charged particle does not experience any force. However, since

$$H = \frac{1}{2m}\left(\vec{p} - \frac{e}{c}\vec{A}\right)^2, \tag{19.8}$$

the wavefunction takes the form

$$\psi = \psi^{(0)} \exp\left[\frac{ie}{\hbar c}\oint d\vec{l}\cdot\vec{A}\right], \tag{19.9}$$

where the line integral is over a closed path of the particle enclosing the solenoid and the phase does not affect any measurement performed on the beam. However, the two beams interfere as schematically shown in Fig. 19.2.

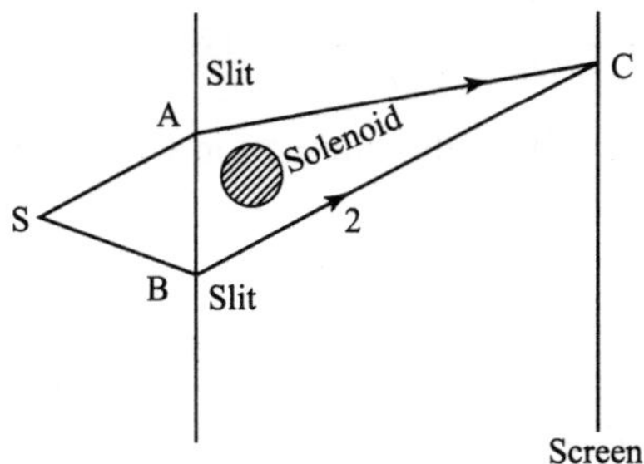

Fig. 19.2 *Schematic display of Bohm–Aharanov experiment*

The wavefunctions for the two beams are

$$\psi_1 = \psi^{(0)} \exp\left[\frac{-ie}{\hbar c}\int_{\text{SAC}} d\vec{l}\cdot\vec{A}\right], \tag{19.10}$$

$$\psi_2 = \psi^{(0)} \exp\left[\frac{-ie}{\hbar c}\int_{\text{SBC}} d\vec{l}\cdot\vec{A}\right], \tag{19.11}$$

where $\psi^{(0)}$ is the wavefunction at S. The phase difference Φ between the two beams at C then becomes

$$\begin{aligned}\Phi &= \frac{e}{\hbar c}\left(\int_{\text{SAC}} - \int_{\text{SDC}}\right) d\vec{l}\cdot\vec{A} \\ &= \frac{e}{\hbar c}\left(\int_{\text{SAC}} + \int_{\text{CBS}}\right) d\vec{l}\cdot\vec{A} = \frac{e}{\hbar c}\int_{\text{SACBS}} d\vec{l}\cdot\vec{A} \\ &= \frac{e}{\hbar c}\oint d\vec{l}\cdot\vec{A}.\end{aligned} \tag{19.12}$$

Applying Stokes theorem,

$$\Phi = \frac{e}{\hbar c}\oint d\vec{S}\cdot\vec{\nabla}\times\vec{A} = \frac{e}{\hbar c}\oint d\vec{S}\cdot\vec{B}, \tag{19.13}$$

where the surface is to circum-enclose the cross-sectional area of the solenoid. This tells us that the fringe system would shift as the magnetic field is varied. This has been verified experimentally.

19.4 Magnetic Dipole in the Field of a Line Charge

Aharanov–Casher effect

The Hamiltonian of a Dirac electron is an external electric field $\vec{E}$, which when expanded in powers of $\frac{1}{c}$ yields (ignoring the Darwin term)

$$H = \frac{p^2}{2m} + eV - \frac{p^4}{8m^3c^2} - \frac{e}{4m^2c^2}\vec{\sigma} \cdot (\vec{E} \times \vec{p}), \tag{19.14}$$

where the last term representing the interaction with $\vec{E}$ can be written as

$$\frac{-e}{2mc}(\vec{A} \cdot \vec{p}) \tag{19.15}$$

with

$$\vec{A} = -\frac{\vec{\mu} \times \vec{E}}{c}, \quad \vec{\mu} = \frac{e\vec{\sigma}}{2mc}. \tag{19.16}$$

For a straight-line charge of linear density ρ, the electric field $\vec{E}$ is at right angles to the wire and has a magnitude

$$E = 4\pi\rho. \tag{19.17}$$

If we consider the diffraction of a neutral particle such as a neutron (which has a magnetic moment) around a wire schematically the same as the Bohm–Aharanov experiment with the electron replaced by a neutron and the solenoid by a straight-line charge, then from (19.16) and (19.17) we have,

$$\oint \vec{A} \cdot d\vec{l} = \pm\frac{4\pi\mu\rho}{\hbar c}, \tag{19.18}$$

where the $\pm$ sign corresponds to neutron spin up or down with respect to the plane of motion of the neutron. Hence the phase difference between the two paths on the screen is

$$\Phi = \pm\frac{4\pi\mu\rho}{\hbar c}. \tag{19.19}$$

The experiment to observe the Aharanov–Casher effect was performed using a neutron interferometer at the University of Missouri, Columbia.

19.5 Phase of the Spin Wavefunction

The spin wavefunction $\psi(s)$ transforms under a rotation by an angle ϕ about the z-axis as

$$\psi(s, \phi) = e^{-is_z\phi/\hbar}\psi(s, 0). \tag{19.20}$$

For a spin-$\frac{1}{2}$ particle one would therefore get, under rotation of 2π,

$$\psi(s, 2\pi) = e^{-i\pi}\psi(s, 0) = -\psi(s, 0), \tag{19.21}$$

$$\psi(s, 4\pi) = \psi(s, 0),$$

i.e., the sign of the wavefunction changes on rotation by an angle of 2π and returns to the original for 4π. This is in contrast to space wavefunction which does not change sign under 2π rotation. In order to test this experimentally, the neutron (which has a $\frac{1}{2}$ spin) in one of its paths in a neutron interferometer is subjected to a uniform magnetic field as shown in the Fig. 19.3, which introduces an extra phase $\delta\Phi$:

$$\delta\Phi = \frac{\omega T}{2}, \quad \omega = \frac{eB}{mc}g_n$$

where $T =$ the time spent by the neutron in the magnetic field and $g_n = -1.91$ is the neutron's g-factor. If the strength of the magnetic field is varied keeping T fixed, constructive interference should give place to destructive interference for $\delta\Phi = 2\pi$; but the experiment shows that this happens when $\delta\Phi = 4\pi$, thereby confirming the theoretical result.

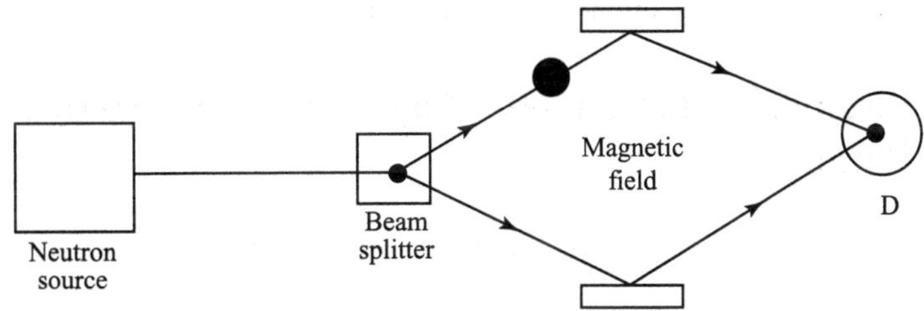

Fig. 19.3 *Phase of spin $\frac{1}{2}$ charged particle wavefunction in a magnetic field*

19.6 Neutral Particle in Earth's Gravitational Field

The potential energy of a particle of mass m in Earth's gravitational field is

$$V(t) = mgz(t), \tag{19.22}$$

where $z(t)$ is the height at which the particle is travelling and g is the acceleration due to gravity. Substituting the above in (19.3) we get

$$\psi = \psi^{(0)} \exp\left(-\frac{img}{\hbar}\int dtz(t)\right). \tag{19.23}$$

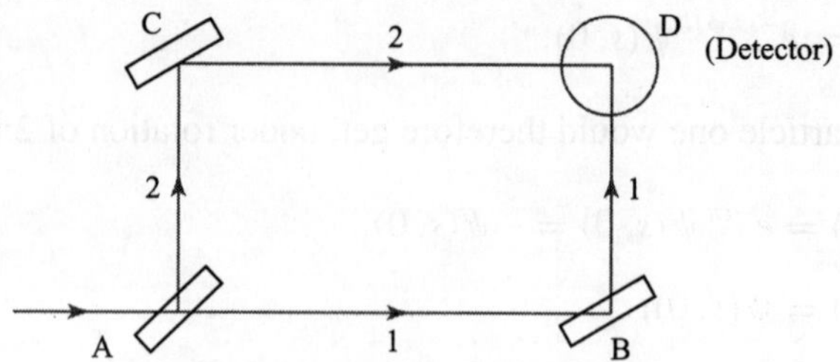

Fig. 19.4 *Neutron interferometer*

If the particle is made to traverse a path as depicted in Fig. 19.4, the phase difference between the two paths, 1 and 2, works out to

$$\Phi = \frac{mg}{\hbar}\int dt\,[z_{CD}(t) - z_{AB}(t)] = \frac{mgl}{\hbar}(t_B - t_A)$$

$$= \frac{mglL}{\hbar v} = \frac{mgS}{\hbar v} = \frac{m^2 gS\lambda}{\hbar^2}. \tag{19.24}$$

Here $v = \frac{2\pi\hbar}{m\lambda}$ is the horizontal velocity of the particle along AB and CD, λ being its de Broglie wavelength. If the figure ABCD is rotated about a horizontal axis, the effective area S would change sinusoidally and so would the phase Φ, giving rise to interference at D. Neutron was used as the neutral particle in the actual experiment by Colella, Overhauser and Werner.

19.7 Phase of Neutron Wavefunction due to Earth's Rotation

A neutral particle like the neutron on Earth's surface is acted upon by the Coriolis force

$$\vec{F}_c = 2m\,\vec{v} \times \vec{\Omega}, \tag{19.25a}$$

where $\vec{\Omega}$ is the angular velocity of rotation of the Earth. In order to obtain its effect on the phase of the neutron, we bring out its similarity with the motion of a charged particle in a magnetic field where the Lorentz force is

$$F_L = \frac{e}{c}\,\vec{v} \times \vec{B} \tag{19.25b}$$

and the canonical momentum is

$$\vec{p} = m\dot{\vec{r}} + \frac{e\vec{A}}{c} = m\dot{\vec{r}} + \frac{e}{2c}\vec{r} \times \vec{B}. \tag{19.26}$$

By comparing (19.25a) and (19.25b), we find for the present case

$$\vec{p} = m\dot{\vec{r}} + m\vec{r} \times \vec{\Omega} = m\dot{\vec{r}} + 2m\vec{V}, \tag{19.27}$$

$$\vec{\Omega} = \vec{\nabla} \times \vec{V},$$

where $\vec{V}$ is the analog of the vector potential $\vec{A}$. A comparison with (19.12) gives

$$\Phi = \frac{2m}{\hbar}\oint d\vec{l} \cdot \vec{V} = \frac{2m}{\hbar}\int d\vec{S} \cdot \vec{\Omega}. \tag{19.28}$$

The experimental arrangement to observe the interference resulting from the phase difference between two split neutron beams is schematically given in Fig. 19.5.

Here θ = the latitude place of observation,

φ = the angle between $\vec{\Omega}$ and $\vec{S}$, which has direction normal to the area enclosed by the paths and the magnitude S.

A simple calculation gives

$$\Phi = \frac{4\pi m}{\hbar} \sin\theta \sin\varphi. \tag{19.29}$$

The phase difference is measured by the coincidence between the neutron counters C_1 and C_2. The experiment performed by Colella et al is qualitatively given

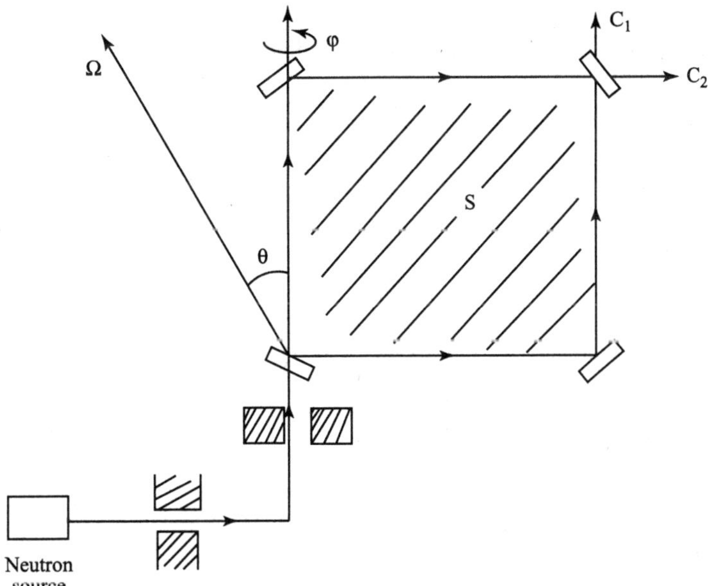

Fig. 19.5 *Interference between neutron beams in Coriolis force field*

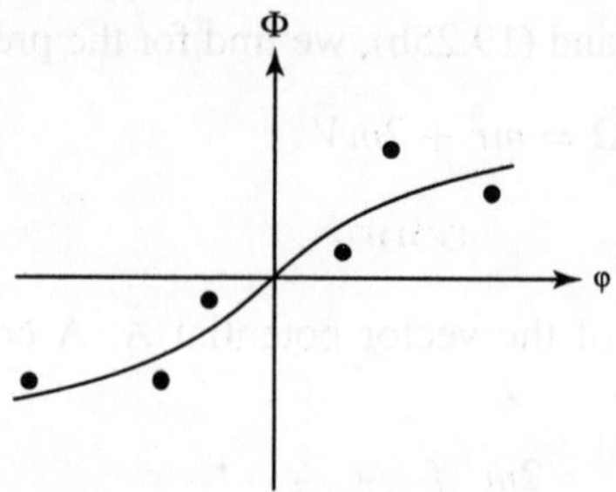

Fig. 19.6 *Qualitative plot of phase of neutron*

in Fig. 19.6, where the solid curve is the theoretical prediction as per (12.29) and the dots are the experimental points.

19.8 Charged Particle in the Field of a Magnetic Charge

The phase of the wavefunction of a charged particle in the static magnetic field of a long and thin solenoid has been discussed in section 4, according to which

$$\Phi = \frac{e}{\hbar c} \int d\vec{l} \cdot \vec{A} \tag{19.30}$$

with

$$\vec{B} = \vec{\nabla} \times \vec{A}. \tag{19.31}$$

The magnetic field due to a point magnetic charge

$$\vec{B} = \frac{g\vec{r}}{r^3} \tag{19.32}$$

can also be expressed as the curl of a vector potential $\vec{A}$,

$$\vec{B} = \vec{\nabla} \times \vec{A},$$

with

$$A_1 = -\frac{gx_2}{r(r+x_3)}, \quad A_2 = \frac{gx_1}{r(r+x_3)}, \quad A_3 = 0. \tag{19.33}$$

It will be seen that for such a field

$$\vec{\nabla} \cdot \vec{B} = 0$$

except at $x_1 = x_2 = 0$ and $x_3 = -r$, i.e. along a line on the entire negative x_3 axis extending from 0 to ∞, where it is singular. This line of singularity is known as the Dirac string.

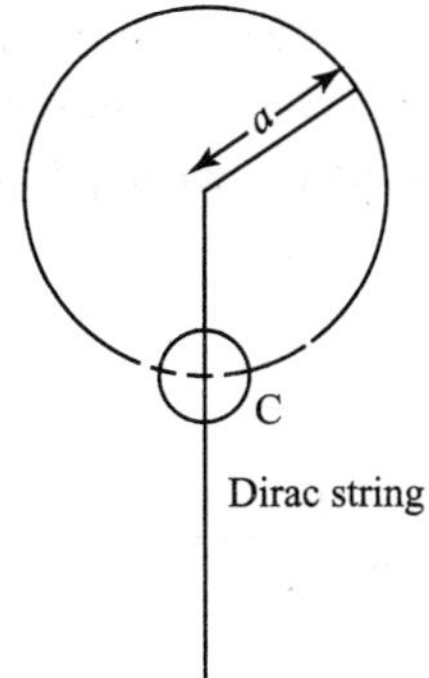

Fig. 19.7 *Contour around Dirac string due to a monopole*

If we construct a sphere of radius a with the magnetic charge at its centre and pierce a thin hole as shown in Fig. 19.7, the phase can be expressed as an integral around the infinitesimally small circle. Then,

$$\Phi = \frac{e}{\hbar c} \oint d\vec{l} \cdot \vec{A} = \frac{e}{\hbar c} \int d\vec{S} \cdot \vec{B}. \tag{19.34}$$

Substitution of (19.32) in (19.34) gives

$$\Phi = \frac{4\pi eg}{\hbar c}. \tag{19.35}$$

The condition of single-valuedness of the wavefunction requires

$$\Phi = 2\pi n,$$

where n is a positive integer which when substituted in (19.35) gives,

$$\frac{eg}{\hbar c} = \frac{n}{2}; \tag{19.36}$$

this is the Dirac's quantization condition.

19.9 Phase of Coherent Macroscopic Quantum System

As mentioned in chapter 2, section 1, in the original formulation of wave mechanics by Schrödinger,

$$\rho = \psi\psi^* \quad \text{and} \quad \vec{j} = \frac{\hbar}{2im}\left[\psi^*(\vec{\nabla}\psi) - (\vec{\nabla}\psi^*)\psi\right] \tag{19.37}$$

were taken as particle charge and current densities (apart from a factor e). But it soon became clear that there would be radiation from the current of the charged particle and Max Born interpreted these as probability densities.

Schrödinger's original interpretation would hold for a system where a very large number of particles occupy a single quantum state which can happen if the particles are bosons. For such a system we can have

$$\psi = \rho^{1/2} \cdot e^{i\Phi}, \tag{19.38}$$

so that

$$\vec{j} = e\rho\vec{v} = \frac{e\hbar}{2im}\psi^* \overleftrightarrow{\nabla} \psi = \frac{\hbar\rho}{m}\vec{\nabla}\Phi, \tag{19.39}$$

$$\vec{p} = m\vec{v} = \hbar\vec{\nabla}\Phi. \tag{19.40}$$

In an external electromagnetic field,

$$\vec{p} = \hbar\vec{\nabla}\Phi - \frac{c\vec{A}}{c},$$

$$\vec{j} = \frac{\hbar\rho}{m}(\nabla\Phi - e \cdot \vec{A}/\hbar c), \tag{19.41}$$

where e is the charge of the particles constituting the macroscopic state.

We now apply these considerations to a superconducting cylinder as depicted in Fig. 19.8 where the flux lines have been shown. Since inside a superconductor

$$\vec{j} = 0, \tag{19.42}$$

we get from (19.41)

$$\vec{\nabla}\Phi = \frac{e\vec{A}}{\hbar c}, \tag{19.43}$$

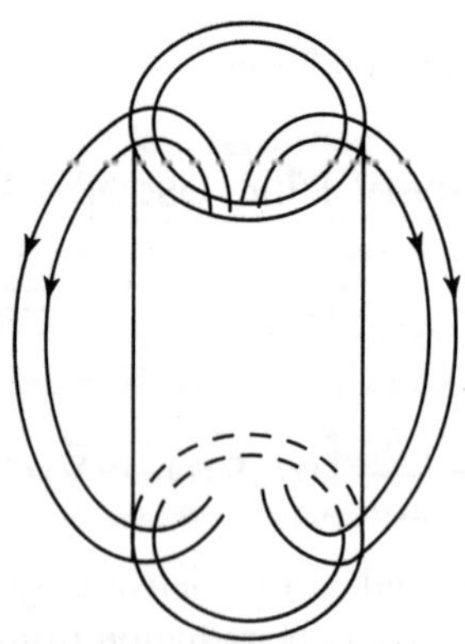

Fig. 19.8 *Magnetic lines of force due to a superconducting cylinder*

and hence

$$\Phi = \frac{e}{\hbar c}\int d\vec{l}\cdot\vec{A} = \frac{e}{\hbar c}\int d\vec{S}\cdot\vec{B} = \frac{eF}{\hbar c}, \tag{19.44}$$

where F = the magnetic flux. Since the wavefunction has to be single-valued,

$$\Phi = 2\pi n \tag{19.45}$$

where n is an integer. Form (19.44) and (19.45) we get

$$F = \frac{2\pi n\hbar}{e} = 4n \times 10^{-7}\ \text{Gauss} \times \text{cm}, \tag{19.46}$$

which means that the flux trapped in the superconducting cylinder is quantized. This relation was obtained in 1950 by F. London; the experiments conducted by Deaver and Fairbank agrees with

$$F = \frac{\pi n\hbar}{e} = 2n \times 10^{-7}\ \text{Gauss} \times \text{cm}, \tag{19.47}$$

which differs from (19.46) by a factor of 2. This is because the particles in the superconducting state are Cooper pairs whose charge is $2e$ rather than e. In the experiment, a copper wire of diameter 1.3×10^{-3}cm covered with tin was put in an external magnetic field and the temperature lowered below 3.8°K at which tin becomes a superconductor. The magnetic field was then switched off and the magnetic moment of the wire measured. It was found that the flux trapped in the superconductor is quantized according the theoretical formula (19.47).

19.10 Topological Phase

The phase $\Phi = \frac{e}{\hbar c}\int d\vec{l}\cdot\vec{A}$ in most of the examples discussed in the preceding sections is not the same as $Et/\hbar$ ordinarily encountered in quantum mechanics. It depends on the geometry of the system and is therefore called geometrical or topological phase. Although known for a long time and occurring in diverse physical problems illustrated in the previous sections, it was M.V. Berry who made a systematic investigation and showed that the standard adiabatic theorem fails in all these cases. He obtained the general result

$$\psi_n(T) = e^{i\gamma_n} e^{-i\int_0^T dt\, E_n(\lambda(t))} \psi_n(0) \tag{19.48}$$

when the Hamiltonian is a function of a slowly varying time-dependent parameter. The phase γ_n is independent of the dynamics and is solely determined by the topology of the λ-space.

In order to obtain a general expression for γ_n, we substitute (19.48) in the equation ($\hbar = 1$)

$$H\psi_n = i\frac{\partial \psi_n}{\partial t}, \quad H \equiv H(\lambda(t)), \tag{19.49}$$

and obtain

$$\gamma_n(t) = i\langle \psi_n(t)|\dot{\psi}_n(t)\rangle, \tag{19.50}$$

from which we get

$$\gamma_n(T) = \int_0^T dt\dot{\gamma}_n(t) = i\int_0^T dt\langle \psi_n(t)|\dot{\psi}_n(t)\rangle. \tag{19.51}$$

This can be put in the form an integral over the λ-space:

$$\gamma_n(t) = \oint d\vec{\lambda}\langle \psi_n(t)|\vec{\nabla}_\lambda \psi_n(t)\rangle = \oint d\vec{\lambda}\cdot \vec{A}_n(\lambda), \tag{19.52}$$

where

$$\vec{A}(\lambda) = i\langle \psi_n|\vec{\nabla}_\lambda \psi_n\rangle$$

is a 'vector potential' in the parameter space $\vec{\lambda}$. By Stoke's theorem, (19.52) can be written as

$$\begin{aligned}
\gamma_n &= i\int d\vec{a}.\vec{\nabla}_\lambda \times \langle \psi_n|\vec{\nabla}_\lambda \psi_n\rangle = i\int da_i \epsilon_{ijk}\frac{\partial}{\partial \lambda_j}\left\langle \psi_n|\frac{\partial \psi_n}{\partial \lambda_k}\right\rangle \\
&= i\int da_i \epsilon_{ijk}\left\langle \frac{\partial \psi_n}{\partial \lambda_j}|\frac{\partial \psi_n}{\partial \lambda_k}\right\rangle = i\epsilon_{ijk}\int da_i \sum_m \left\langle \frac{\partial \psi}{\partial \lambda_j}|\psi_m\right\rangle\left\langle \psi_m|\frac{\partial \psi_n}{\partial \lambda_k}\right\rangle \\
&= i\epsilon_{ijk}\int da_i \sum_m \frac{\langle \psi_n|\frac{\partial H}{\partial \lambda_j}|\psi_m\rangle\langle \psi_m|\frac{\partial H}{\partial \lambda_k}|\psi_n\rangle}{(E-E_m)^2}.
\end{aligned} \tag{19.53}$$

Problems

1. For a spin-$\frac{1}{2}$ particle coupled to a time-dependent magnetic field $B(t)$, show that the Berry phase for the two energy eigenvalues are

$$\gamma_\pm = \pm\frac{1}{2}\int d\vec{a}\cdot \frac{\vec{B}}{B^2}. \tag{19.54}$$

Hint: Take $\vec{B}$ as the parameter $\vec{\lambda}$ and obtain the energy eigenvalues and the eigenfunctions of the Hamiltonian $H = \frac{1}{2}\vec{\sigma}\cdot \vec{B}$ and obtain $\gamma_\pm$ using (15.52) and (15.53).

Bibliography

[1] R. Collella, A. Ovewrhauser and S.A. Werner, *Phys. Rev. Lett.*, **34**, 1472 (1975); **42**, 1103 (1979).

[2] D. Bohm and Y. Aharanov, *Phys. Rev.*, **115**, 485 (1959).

[3] Y. Ahranov and A. Casher, *Phys. Rev. Lett.*, **53**, 319 (1984).

[4] P.A.M. Dirac, *Proc. Roy. Soc.*, A**133**, 60 (1931).

[5] R.G. Chambers, *Phys. Rev. Lett.*, **5**, 3 (1960).

[6] H. Kaiser et.al., *Physica*, **151B**, 68 (1988).

[7] F. London, *Superfluids*, Vol. I., John Wiley & Sons, New York, (1950).

[8] B.S. Deaver Jr. and W.M. Fairbank, *Phys. Rev. Lett.*, **7**, 51 (1961).

[9] M.V. Berry, *Proc. Roy. Soc.*, **A392**, 45 (1984).

[10] P.W. Anderson, *Rev. Mod. Phys.*, **38**, 298 (1966).

[11] R.P. Feynman, R.B. Leighton and M. Sands, *The Feynman Lectures*, Vol. III, Addison Wesley Publishing Co. Inc., Reading, Massachusetts (1965).

[12] J.J. Sakurai, *Modern Quantum Mechanics*, Addison-Wesley Publishing Company Inc. (1985).

20 Lagrangian Formulation of Quantum Mechanics

20.1 Dirac's Transformation Theory Approach

Quantum mechanics as formulated by Heisenberg and Schrödinger using different approaches and found to be equivalent is based on the Hamiltonian formalism of classical mechanics. In 1933, Dirac gave a Lagrangian formulation of quantum mechanics which is based on the Hamilton's principal function. He noted that in classical mechanics the dynamical variables q and p at a time t_0 are connected to the these variables at any other time by a contact (canonical) transformation generated by the 'action function' S, which is the time integral of the Lagrangian taken between these two times. Usually, a contact transformation consists of

$$q, p \longrightarrow Q_i(q, p), P_i(q, p),$$

$$\text{where} \quad P = -\frac{\partial S}{\partial Q}, \; p = \frac{\partial S}{\partial q}, \tag{20.1}$$

with

$$S = S(q, Q).$$

If one takes $q = q(t), p = p(t), Q = q(t_0), P = p(t_0)$, then (20.1) has the form

$$p(t_0) = -\frac{\partial S}{\partial q(t_0)}, \; p(t) = \frac{\partial S}{\partial q(t)},$$

with

$$S = \int_{t_0}^{t} dt' L(t'), \tag{20.2}$$

which is Hamilton's first principal function; $L(t)$ is Lagrangian. The main hurdle in going over to quantum mechanics is that the Lagrange equation involves partial derivatives of the Lagrangian with respect to the coordinates and velocities which can be done in terms of commutators with the Hamiltonian and one is back to square one, i.e., the Hamiltonian theory. Dirac overcame this problem through what has come to be known as 'transformation theory', in which the key element is the transformation function $\langle q|Q\rangle$ which connects the representation in which q is diagonal with the one in which Q is diagonal. In general, an operator α can have matrix elements like $\langle q|\alpha|Q\rangle$ which can be related to both $\langle q|\alpha|q'\rangle$ and $\langle Q|\alpha|Q'\rangle$ with the aid of the transformation function

$$\langle q|\alpha|Q\rangle = \int \langle q|\alpha|q'\rangle dq'\langle q'|Q\rangle = \int \langle q|Q'\rangle dQ'\langle Q'|\alpha|Q\rangle. \tag{20.3}$$

If we put $\alpha = q$, we get, from the above equation,

$$\langle q'|q|Q'\rangle = q'\langle q'|Q'\rangle. \tag{20.4}$$

Similarly, if we put $\alpha = p$, we get

$$\langle q'|p|Q'\rangle = -i\hbar\frac{\partial}{\partial q'}\langle q'|Q'\rangle. \tag{20.5}$$

Following the same procedure, we get with $\alpha = Q$ and $\alpha = P$,

$$\langle q'|Q|Q'\rangle = Q'\langle q'|Q'\rangle, \tag{20.6}$$

$$\langle q'|P|Q'\rangle = i\hbar\frac{\partial}{\partial Q'}\langle q'|Q'\rangle. \tag{20.7}$$

If we take

$$\langle q'|Q'\rangle = \exp\left[\frac{iU}{\hbar}(q'Q')\right], \tag{20.8}$$

and use it in (20.5) and (20.7), we get

$$\langle q'|p|Q'\rangle = \frac{\partial U(q'Q')}{\partial q'}\langle q'|Q'\rangle, \tag{20.9}$$

$$\langle q|P|Q\rangle = -\frac{\partial U(q', Q')}{\partial Q'}\langle q'|Q'\rangle. \tag{20.10}$$

Comparing (20.9) and (20.10) with (20.5) and (20.7),

$$p = \frac{\partial U}{\partial q}, \quad P = -\frac{\partial U}{\partial Q}.$$

These are quantum analog of the classical contact transformation, U being the analog of the classical action S. We now set $q = q(t)$ and $Q = q(t_0)$ in (20.8) for quantum mechanics as was done in the classical case. It is clear that

$$\langle q(t)|q(t_0)\rangle \ \xrightarrow{\text{corresponds to}} \ \exp\left[\frac{i}{\hbar}\int_{t_0}^{t} dt L(t)\right]. \tag{20.11}$$

It we take $t = t_0 + dt$, then we have

$$\langle q(t+dt)|q(t)\rangle \ \xrightarrow{\text{corresponds to}} \ \exp\left[\frac{i}{h}L\,dt\right]. \tag{20.12}$$

This suggests that the Lagrangian should be taken as a function of $q(t)$ and $q(t+dt)$ rather than $q(t)$ and $\dot{q}(t)$.

The correspondence between the left- and the right-hand sides of (20.11) can be examined by breaking up the time interval $(t - t_0)$ into small sections so that

$$\begin{aligned} B(t, t_0) &= e^{\frac{i}{\hbar}\int_{t_0}^{t} dt' L(t')}, \\ &= B(t, t_m)B(t_m, t_{m-1})\cdots B(t_2, t_1)B(t_1, t_0), \end{aligned} \tag{20.13}$$

$$\begin{aligned} \langle q(t)|q(t_0)\rangle = \int &\langle q(t)|q(t_m)\rangle dq(t_m)\langle q(t_m)|q(t_{m-1})\rangle dq(t_{m-1})\cdots \\ &\langle q(t_2)q(t_1)\rangle dq(t_1)\langle q(t_1)|q(t_0)\rangle. \end{aligned} \tag{20.14}$$

It will be seen that while (20.13) has no integration, (20.14) has many. However, on account of (20.8) each transformation function has the form $e^{iF/\hbar}$, where F is a function of all the q's. Because $\hbar$ is small, $F/\hbar$ varies rapidly with q's unless F itself varies little. Thus the contribution to the integrals come from a very small region where F is stationary with respect to small variations in q and there is no contradiction between (20.13) and (20.14); the correspondence (20.11) holds and we have what is known as Dirac's 'action principle' in quantum mechanics.

The square of the transformation function can be interpreted as the probability of q having value $q(t)$ at time t for a state for which an observation of q at an earlier time t_0 is certain to give the result $q(t_0)$. This is a kind of probability of a *path* in contrast to the Hamiltonian quantum mechanics which gives the probability of *position*. In order to obtain the connection of the Lagrangian approach with the Hamiltonian approach, we note that the Schrödinger wave equation can be written in terms of the transformation function $\langle x'|x\rangle$ in the integral form as

$$\psi(x', t+\epsilon) = \int dx (x'|x > \psi(x, t). \tag{20.15}$$

This can be verified as follows:

$$\begin{aligned} \langle x'|x\rangle = \langle x'(t')|x(t)\rangle|_{t'=t+\epsilon} &= \langle x'(t)|U(t+\epsilon, t)|x(t)\rangle \\ &= \exp\left(\frac{i\epsilon}{\hbar}H\right)\langle x'(t)|x(t)\rangle. \end{aligned} \tag{20.16}$$

Substituting (20.16) in (20.15) we get

$$\psi(x', t+\epsilon) = \int dx \langle x'(t)|x(t)\rangle \exp\left(\frac{i\epsilon H}{\hbar}\right) \psi(x, t). \tag{20.17}$$

Since

$$\langle x'(t)|x(t)\rangle = \delta(x - x')$$

and

$$\exp\left(\frac{i\epsilon H}{\hbar}\right) \psi(x, t) = \psi(x, t+\epsilon),$$

equation (20.15) is verified which establishes the desired connection.

20.2 Feynman's Path Integral Approach

We have seen in the previous section that in th Lagrangian approach to quantum mechanics, the dynamics is determined by the transformation function $\langle q(t)|q(t_0)\rangle$, which is interpreted by Dirac as the probability amplitude of the position operator q having a value $q(t)$ at time t for a state for which an observation at an earlier time t_0 is certain to give result $q(t_0)$. This is a probability amplitude for a particle path rather than its position which one encounters in Hamiltonian quantum mechanics. Feynman bases his formulation of quantum mechanics on such paths the basic principles of which can be stated as:

The probability $P(b, a)$ of a particle moving from a point a to b is given by the square modulus of a complex number $K(b, a)$:

$$P(b, a) = |K(b, a)|^2. \tag{20.18}$$

The amplitude of this probability $K(b, a)$ gets contribution from all possible paths from a to b.

$$K(b, a) = \sum_{\text{path}} k e^{iS/\hbar}. \tag{20.19}$$

The paths contribute equally in magnitude, but with the phase determined by the classical action S in units of $\hbar$.

$$S = \int_{\text{path}} dt L(t), \tag{20.20}$$

where $L(t)$ is the classical Lagrangian. The magnitude k is fixed by the formula

$$K(c, a) = \sum_{b} K(c, b) K(b, a). \tag{20.21}$$

The complex function $K(b, a)$ is actually Dirac's transformation function:

$$K(x(t), x'(t')) = \langle x(t)|x'(t')\rangle, \tag{20.22}$$

which, as per previous section can be written as

$$\langle x(t)|x'(t')\rangle \xrightarrow{\text{corresponds to}} \prod_{n=1}^{N} \exp\left[\frac{i}{\hbar}S(x_n t_n, x_{N-1}t_{n-1})\right], \tag{20.23}$$

i.e.,

$$\langle x_n(t_n)|x_{n-1}(t_{n-1})\rangle \xrightarrow{\text{corresponds to}} \exp\left[\frac{i}{\hbar}S(x_n t_n,), x_{N-1}(t_{n-1})\right]. \tag{20.24}$$

Feynman put forth the point that if the time interval $(t_n - t_{n-1}) = n \in\to 0$, the path becomes a straight line. In this case $n \propto \frac{1}{\epsilon}$ but $t_n = t$ is finite. He therefore put the equality

$$\langle x_n(t_n)|x_{n-1}(t_{n-1})\rangle \lim_{\substack{\epsilon\to 0\\ N\to\infty}} \frac{1}{A} \exp\left[\frac{i}{\hbar}(x_n(t_n,), x_{n-1}(t_{n-1}))\right] \tag{20.25}$$

in place of Dirac's relation (20.24). The constant A occurring in this equation depends on the Lagrangian. From this, Feynman obtained

$$\begin{aligned}\langle x(t), x'(t')\rangle &= \lim_{\substack{\epsilon\to 0\\ N\to\infty}} \int dx_1 \int dx_2 \cdots \int dx_N \prod_{n=1}^{N} \langle x_n(t_n)|x_{n-1}(t_{n-1})\rangle,\\ &= \lim_{\substack{\epsilon\to 0\\ N\to\infty}} \int_{\frac{dx_1}{A}} \int \frac{dx_2}{A} \cdots \int \frac{dx_N}{A}\\ &\quad \times \exp\left[\frac{i}{\hbar}S(x_n(t_n,), x_{n-1}(t_{n-1}))\right],\end{aligned} \tag{20.26}$$

where

$$S(x_n(t_n), x_{n-1}(t_{n-1})) = \int_{t_{n-1}}^{t_n} dt' L(x'(t', \dot{x}'(t')). \tag{20.27}$$

Equation (20.26) can be written in shorthand as

$$\langle x(t)|x'(t')\rangle = \int_{x'(t')}^{x(t)} D(x(t)) \exp\frac{i}{\hbar}S(x(t), x'(t')). \tag{20.28}$$

Feynman's formula (20.26) for Dirac's transformation function can be considered as a sum of action over space-time paths. Dirac had considered such splitting of the path as can be seen from the presentation in the previous section, but his final formula was a correspondence, not an equality.

In Feynman's formalism, the transformation function is taken as the wave function

$$\langle x(t)|x'(t')\rangle = \psi(x,t), \tag{20.29}$$

which, although is the probability amplitude for a particle coming from $x'(t')$ to be $x(t)$, has more information than what one would obtain from $\psi(x,t)$. With the definition (20.28), equation (20.21) would become

$$\psi(x,t) = \int dx''\langle x(t)|x''(t'')\rangle\psi(x'',t''). \tag{20.30}$$

Physically, it tells us that the amplitude $\psi(xt)$ for the particle to arrive at $x(t)$ is the sum (integral) over all possible values $x''(t'')$ to arrive at $x''(t'')$ is $\psi(x''t'')$ multiplied by the transformation function $\langle x(t)|x''(t'')\rangle$.

20.3 Dirac–Feynman Action Principle

Combining (20.25) and (20.30) of the foregoing section we get

$$\psi(x_{n+1}, t+\epsilon) = \int \frac{dx_n}{A} \exp\left[\frac{i}{\hbar}S(x_{n+1}, x_n)\right]\psi(x_n, t), \tag{20.31}$$

which is known as 'Feynman equation' for the wavefunction. It tells that if the amplitude of the wave ψ is known on a given surface consisting of all x_n at a given time t, its value at a nearby point at a time $(t+\epsilon)$ is sum of contributions from all points on the surface at time t. Each contribution is delayed by an amount proportional to the action it would require to get from the surface to the point along the path of least action of classical mechanics. In this sense it is an action principle which can be named as Dirac–Feynman action principle. It can also be called the Huygen's principles for matter waves, as Feynman puts it. He also notes that Kirchoffs' corrections to Huygen's principle are not needed since one has the first-order derivative in time.

20.4 Equivalence of Feynman and Schrödinger Equations

The first step in establishing the equivalence is to determine the normalisation constant A occurring in the Feynman equation (20.31). This is most easily done for a particle moving in one dimension under the action of a force for which

$$S(x_{n+1}, x_n) = \frac{m\epsilon}{2}\left(\frac{x_{n+1}-x_n}{\epsilon}\right)^2 - \epsilon V(x_{n+1}), \tag{20.32}$$

which on substitution in (20.31) gives

$$\psi(x_{n+1}, t+\epsilon) = \int \frac{dx_n}{A} \exp\left[\frac{i\epsilon}{\hbar}\left\{\frac{m}{2}\left(\frac{x_{n+1}-x_n}{\epsilon}\right)^2 - V(x_{n+1})\right\}\right]. \tag{20.33}$$

In terms of

$$x = x_{n+1} \qquad \xi = x_{n+1} - x_n,$$

equation (20.33) has the form

$$\psi(x, t+\epsilon) = \int \frac{d\xi}{A} \exp\left[\frac{i\epsilon}{\hbar}\left(\frac{m\xi^2}{2\epsilon} - \epsilon V(x)\right)\right] \psi(x-\xi, t). \tag{20.34}$$

In the limit $\epsilon \to 0$, the term $e^{im\xi^2/2\hbar\epsilon}$ oscillates rapidly giving zero contribution to the integral except for small ξ. We can, therefore, expand $\psi(x-\xi, t)$ in a Taylor series in ξ and obtain

$$\psi(x, t+\epsilon) = e^{\frac{-i\epsilon}{\hbar}V(x)} \int \frac{d\xi}{A} \exp\left(\frac{im\xi^2}{2\hbar\epsilon}\right) \times \left[\psi(x,t) - \xi\frac{\partial\psi}{\partial x} + \frac{\xi^2}{2}\frac{\partial^2\psi}{\partial x^2} + \cdots\right]. \tag{20.35}$$

Performing integration, we get

$$\psi(x, t+\epsilon) + \epsilon\frac{\partial\psi(xt)}{\partial t} = \frac{1}{A}e^{\frac{-i\epsilon V(x)}{\hbar}}\left(\frac{2\pi i\hbar}{m}\right)^{1/2} \times \left[\psi(xt) + \frac{i\epsilon\hbar}{2m}\frac{\partial^2\psi}{\partial x^2} + \cdots\right]. \tag{20.36}$$

The terms of order ϵ on both sides of this equation agree if

$$A = \left(\frac{2\pi i\epsilon\hbar}{m}\right)^{1/2}. \tag{20.37}$$

Further, taking

$$e^{\frac{-i\epsilon V(x)}{\hbar}} = 1 - \frac{i\epsilon}{\hbar}V(x) \tag{20.38}$$

and substituting (20.39) and (20.40) in (20.38) gives

$$i\hbar\frac{\partial\psi}{\partial t} = -\frac{\hbar^2}{2m}\frac{\partial^2\psi}{\hbar}V(x) \tag{20.39}$$

which establishes the equivalence between Feynman and Schrödinger equation.

The equivalence can also be established by showing that for a free particle moving in one dimension, the probability amplitude $K(ab)$ defined in section 20.2 satisfies the Schrödinger equation.

For this case,

$$L = \frac{1}{2}m\dot{x}^2. \tag{20.40}$$

Writing

$$\dot{x}^2 = \left(\frac{x_n - x_{n+1}}{dt}\right)^2 dt = \frac{1}{\epsilon}(x_n - x_{n+1})^2 \tag{20.41}$$

where $dt = \epsilon$, we have, from (20.26),

$$K(a,b) = \lim_{\substack{\epsilon\to 0\\ n\to\infty}} \int \frac{dx_1}{A}\int \frac{dx_2}{A}\cdots\int \frac{dx_{n-1}}{A}\exp\left[\frac{im}{2\epsilon}\sum_{n=1}^{n}(x_n - x_{n-1})^2\right]. \tag{20.42}$$

Substituting A from (20.37) in (20.42) and making use of the identity

$$\int_{-\infty}^{\infty} dx_2 e^{-a(x_1-x_2)^2 - a(x_2-x_3)^2} = \left(\frac{\pi}{a}\right)^{1/2} e^{\frac{-1}{2}(x_1-x_3)^2} \tag{20.43}$$

and carrying out all integrations, the final result comes out to be

$$K(a,b) = \left[\frac{m}{2\pi i\hbar(t_b - t_a)}\right]^{1/2} \exp\left[\frac{im(x_b - x_a)^2}{2(t_b - t_a)}\right] \tag{20.44}$$

which is the Green's function for one-dimensional Schrödinger equation and satisfies the equation

$$-\frac{\hbar^2}{2m}\frac{\partial^2}{\partial x_a^2}K(a,b) = i\hbar\frac{\partial}{\partial t_a}K(a,b). \tag{20.45}$$

20.5 Concluding Remarks

It is instructive to put the four formulations of quantum mechanics for comparison in the same form as has been done by Feynman.

1. Schrödinger wave mechanics

$$\psi(x, t+\epsilon) = -\frac{i\epsilon H}{\hbar}\psi(x,t) \tag{20.46}$$

2. Heisenberg quantum mechanics

$$x(t+\epsilon) = \exp\left[\frac{i\epsilon H}{\hbar}\right] x(t) \exp\left[-\frac{i\epsilon H}{\hbar}\right] \tag{20.47}$$

3. Dirac's Lagrangian quantum mechanics

$$\psi(x', t+\epsilon) = \int dx \langle x'|x\rangle_\epsilon \psi(xt) \tag{20.48}$$

4. Feynman formulation

$$\psi(x_{k+1}, t+\epsilon) = \int \frac{dx_k}{A} \exp\left[\frac{i}{\hbar} S(x_{k+1}, x_k)\right] \psi(x_k, t) \tag{20.49}$$

In this connection, the following remarks are in order.

(a) The wavefunction $\psi(x', t+\epsilon)$ represents a state in a representation in which x' is diagonal while $\psi(x, t)$ represents a state in which x is diagonal. Dirac's transformation function $\langle x'|x\rangle_\epsilon$ relates these two.

(b) Dirac's $\langle x'|x\rangle_\epsilon$ and Feynman's $\frac{1}{A}\exp[\frac{i}{\hbar}S(x', x)]$ are analogous.

(c) Feynman states in his paper in *Reviews of Modern Physics* that 'Dirac's remarks were the starting point of the present development'.

Bibliography

[1] P.A.M. Dirac, Physickalische Zeifschribt der Sowjetunion **3**, 1 (1933) *The Principle of Quantum Mechanics*, Oxford University Press, Fourth edition (1957), section 32; Indian Edition (2004).

[2] R.P. Feynman, *Rev. Mod. Phys.*, **20**, 3678 (1948).

[3] F.S. Levin, *An Introduction to Quantum Theory*, Cambridge University Press (2002).

21 Paradoxes in Quantum Mechanics

21.1 Schrödinger's Cat Paradox

In spite of overwhelming success of quantum mechanics in the understanding the atomic, molecular and nuclear phenomena, there remains several questions that puzzled even the founders of the theory. One such question deals with the postulate that any normalisable wavefunction can be expanded in terms of a complete set of orthonormal eigenfunctions, known as the principle of superposition. And according to the Copenhagen interpretation, before measurement, the system is simultaneously in all those states. A measurement forces the system to collapse to the particular state whose property is measured. Thus the quantum mechanical system has no properties until a measurement is performed on it. This is quite different from classical mechanics and quite baffling to common sense based on experience with classical objects. Schrödinger highlighted this by devising a gedanken experiment depicted in Fig. 21.1

It consists of a cat in a box with a cyanide capsule over its head which can be hit by a hammer to release the poison for killing it. The hammer is controlled by a lever which operates when a single electron is incident upon it. This electron comes from a half silvered mirror which has a definite probability of passing

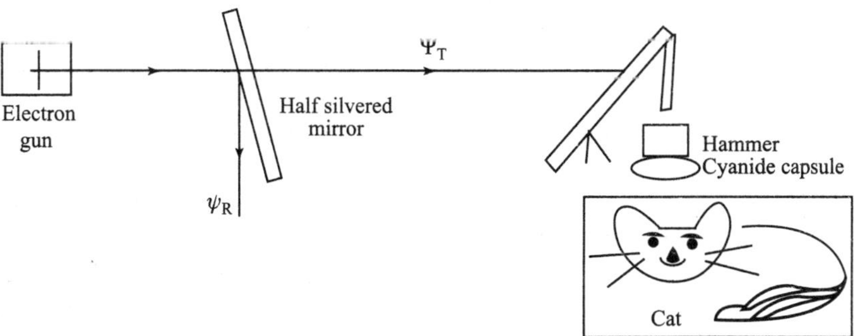

Fig. 21.1 *Schematic presentation of Schrödinger's cat experiment*

through it undeflected and getting reflected. Therefore the state of the electron, say, at time $t = 0$ will be

$$|\phi_e\rangle = |\phi_R\rangle + |\phi_T\rangle, \tag{21.1}$$

and if the cat is in the state $|\psi\rangle$ at this time, then combined state of the electron and the cat will be

$$|\Psi(0)\rangle = |\psi\phi_T\rangle + |\psi\phi_R\rangle. \tag{21.2}$$

At a later time t,

$$\Psi(t) = U(t)|\psi\phi_T\rangle + U(t)|\psi\phi_R\rangle, \tag{21.3}$$

where $U(t) = e^{-Et/\hbar}$ is the time development operator. Since the reflected electron cannot kill the cat

$$U(t)|\psi\phi_R\rangle = |\Psi_A(t)\rangle, \tag{21.4}$$

(subscript A standing for the cat alive) for the same reason

$$U(t)|\psi\Phi_T\rangle = |\Psi_D(t)\rangle, \tag{21.5}$$

subscript D standing for the cat dead. Substituting (21.4) and (21.5) in (21.3) we have

$$|\Psi(t)\rangle = |\Psi_A(t)\rangle + |\Psi_D(t)\rangle, \tag{21.6}$$

i.e. the cat is both alive and dead. If we now make a measurement, i.e., observe the cat, we shall find that it either alive or dead. The state $|\Psi\rangle$ collapses to either of the two states $|\Psi_A\rangle$ or $|\Psi_D\rangle$. The paradox is: how can the cat be both alive and dead? How can a mere look at the cat force it to be either alive or dead? The paradox is resolved by saying that a macroscopic object like cat cannot be a quantum object.

It may be noted here that the state $|\Psi(t)\rangle$ as given in (21.3) is not a simple product of the state of the electron and the state of the cat, but a linear combination of such products. In this state the electron and the cat cannot be separated. Such states are called 'entangled states'.

It may be further noted in this connection that atomic physicists have been able to create in the laboratory atomic states which can simultaneously be in two states. These may be called the atomic Schrödinger cat states or the mesoscopic Schrödinger cat states.

21.2 Einstein–Podolosky–Rosen (EPR) Paradox

Although he pioneered the quantum physics era with his theory of photoelectric effect, Einstein was not happy about the probabilistic nature of quantum mechanics which he expressed by saying that "God does not play dice...". As far back as 1927 he pointed out by a simple argument how it gives rise to action-at-a distance behaviour which runs as follows: Consider a single electron passing through a hole and getting observed on a fluorescent screen at a finite distance as depicted in Fig. 21.2.

Just before causing a scintillation at a point on the screen, the particle has certain probability $|\psi(\vec{r})|^2$ to be at all points $\vec{r}$ on it. When it is actually observed at the point P, it ceases to be anywhere else; according to Copenhagen interpretation the wave function suddenly collapses to that point. Einstein argued that there is a peculiar action-at-a distance that prevents the electron from being observed at other points on the screen. This he called the non-locality of quantum mechanics and considered it as arising from certain inadequacy or incompleteness of quantum mechanics. This was more dramatically illustrated by him and his co-authors Podolsky and Rosen (EPR) through the decay of a spin-zero particle into two photons (which could also be two spin-$\frac{1}{2}$ particles). In the rest system of the spin-zero particle, the two photons γ_1 and γ_2 travel in opposite directions on account of momentum conservation and both have right-circular polarisation on account of angular momentum conservation as shown in Fig. 21.3

Both photons can also have left-circular polarisation. The two-photon quantum mechanical state can therefore be written as

$$|\phi\rangle = \frac{1}{\sqrt{2}}[|R_1R_2\rangle - |L_1L_2\rangle]; \tag{21.7}$$

the minus sign is required to make $|\phi\rangle$ to be a zero-spin state. It follows that when observation is made and γ_1 is found to have right-circular polarisation (R_1), γ_2 will also have right-circular polarisation and the same holds for left-circular

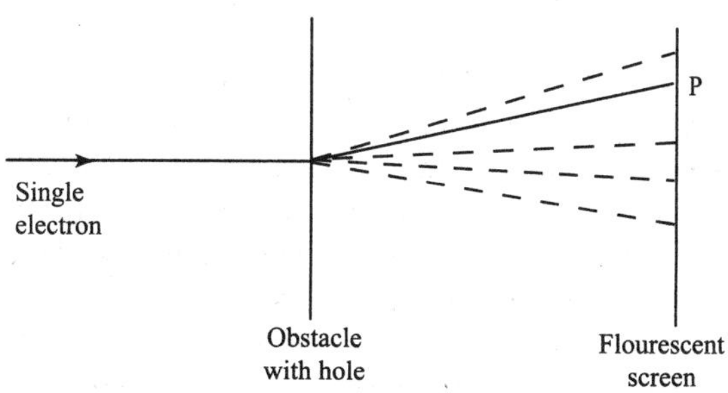

Fig. 21.2 *Einstein's action-at-a-distance paradox*

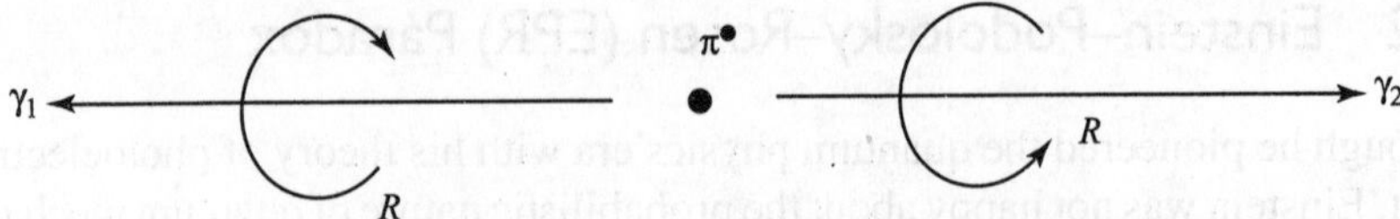

Fig. 21.3 *Polarization of photons from spin-zero particle decay*

polarisations. Further, if we express circular polarisation in terms of plane polarisation X and Y at right angles,

$$R = X + iY,\ L = X - iY, \tag{21.8}$$

then the two-photon state can be expressed in terms of X and Y as

$$|\phi\rangle = \frac{i}{\sqrt{2}}[|X_1 Y_2\rangle + |Y_1 X_2\rangle], \tag{21.9}$$

from which it follows that if γ_1 has X polarisation, γ_2 has Y polarisation and vice versa. EPR argued that there must be an action-at-a-distance for γ_2 to jump to R when γ_1 jumps to R, and to jump to L when γ_1 jumps to L. In other words, they said, quantum mechanics is non-local. That there is some action-at-a distance meaning that there is something that is unknown or hidden, led them to conclude that quantum mechanics is incomplete. This led David Bohm to propose his hidden variable theory which we have noted in chapter 2.

21.3 Bell Inequality for Hidden Variable Theory

J.S. Bell in 1964 obtained an inequality for the probabilities of spin-zero particle to decay into two spin-half particles in different directions within the framework of local hidden variable theories and showed that this inequality is violated by the quantum mechanical probabilities. The derivation of the inequality is as follows:

Consider two spin-half particles in a singlet state:

$$|\phi\rangle = \frac{1}{\sqrt{2}}[|\alpha(1)\rangle|\beta(2)\rangle - |\alpha(2)\rangle|\beta(1)\rangle], \tag{21.10}$$

where $|\alpha(1)\rangle$ is the spin-up state vector for particle no.1 and $|\beta(1)\rangle$ is the spin down state vector of the same particle with same notation for particle no.2. Let A be the outcome of the spin projection of particle no. 1 in the direction of unit vector $\vec{a}$, B be that of particle no.2 in the direction of unit vector $\vec{b}$. A will depend on the hidden variable parameter λ and $\vec{a}$. Because of the correlation arising from entanglement, A could also depend on $\vec{b}$ and similarly, B would depend on $\vec{a}$. Since this kind of dependence of both A and B on $\vec{a}$ and $\vec{b}$ would introduce non-local action at a distance between the two particles, Bell introduced the locality requirement which is

$$A \equiv A(\lambda\vec{a}), \qquad B \equiv B(\lambda\vec{b}). \tag{21.11}$$

In this local hidden variable theory, the correlated probability of spin projection of particle no. 1, i.e. $\vec{\sigma}(1)\cdot\vec{a}$ and spin projection $\vec{\sigma}(2)\cdot\vec{b}$ of particle no. 2, denoted by $P(\vec{a},\vec{b})$ is given by

$$P(\vec{a},\vec{b}) = \int d\lambda\rho(\lambda)A(\lambda,\vec{a})B(\lambda,\vec{b}), \tag{21.12}$$

where $\rho(\lambda)$ is a weight factor such that

$$\int d\lambda\ \rho(\lambda) = 1. \tag{21.13}$$

The requirement of minimum value $P(\vec{a},\vec{b})$ to be (-1) are

$$P(\vec{a},\vec{a}) = -1, \tag{21.14}$$

$$B(\vec{a},\lambda) = -A(\vec{a},\lambda). \tag{21.15}$$

Using (21.15) in (21.12) gives

$$P(\vec{a},\vec{b}) = -\int d\lambda\rho(\lambda)A(\vec{a},\lambda)A(\vec{b},\lambda). \tag{21.16}$$

Using this equation it is possible to obtain

$$|P(\vec{a},\vec{b}) - P(\vec{a},\vec{c})| \leq [1 + P(\vec{b},\vec{c})], \tag{21.17}$$

which is the Bell's inequality for any hidden variable theory.

In quantum mechanics,

$$\begin{aligned} P(\vec{a},\vec{b}) &= \langle\phi|(\vec{\sigma}(1)\cdot\vec{a})(\vec{\sigma}(2)\cdot\vec{b})|\phi\rangle \\ &= \frac{1}{2}a_i b_j\langle[\alpha(1)\beta(2) - \alpha(2)\beta(1)]|(\sigma_i(1)\sigma_j(2))|[\alpha(1)\beta(2) \\ &\quad - \alpha(2)\beta(1)]\rangle. \end{aligned} \tag{21.18}$$

Calculation of this can be performed using σ_i matrices and

$$|\alpha\rangle = \begin{pmatrix}1\\0\end{pmatrix}, \qquad |\beta\rangle = \begin{pmatrix}0\\1\end{pmatrix}, \tag{21.19}$$

which finally gives

$$P(\vec{a},\vec{b}) = -\vec{a}\cdot\vec{b}. \tag{21.20}$$

It is found that this quantum expression does not always satisfy the Bell inequality (21.17). For instance, when

$$\vec{a} = \frac{\vec{b} - \vec{c}}{|\vec{b} - \vec{c}|}, \quad \text{with} \quad \vec{b} \cdot \vec{c} = 0, \tag{21.21}$$

the quantum value (21.20) when substituted in the inequality (2.17) leads to

$$\sqrt{2} \leq 1 \tag{21.22}$$

which is absurd. Correlations obtained experimentally are in agreement with quantum mechanics and also violate Bell's inequality. It is thus seen that results of quantum mechanics cannot be obtained from a hidden variable theory; the two are different.

Bibliography

[1] E. Schrödinger, *Naturwissenschaften*, **23**, 807 (1935); **23**, 923 (1935); **23**, 844 (1935).
[2] Michael W. Noel and C.R. Stroud, *Jr. Phys. Rev. Lett.*, **77**, 1913 (1996).
[3] C. Monroe, et. al., *Science*, **272**, 1131 (1996).
[4] A. Einstein, B. Podolsky and N. Rosen, *Phys. Rev.*, **47**, 777 (1935).
[5] J.S. Bell, *Physics*, **1**, 195 (1964).
[6] F.S. Levin, *An Introduction to Quantum Theory*, Cambridge University Press (2002).
[7] Stuart J. Freedman and John F. Clauser, *Phys. Rev. Lett.*, **28**, 938 (1972).
[8] Edward S. Fry and Randal C. Thompson, *Phys. Rev. Lett.*, **37**, 465 (1976).
[9] A. Aspect, P. Grangies and G. Roges, *Phys. Rev. Lett.*, **47**, 460 (1981).
[10] A. Aspect, J. Dalbart and R. Roger, *Phys. Rev. Lett.*, **49**, 1804 (1982).

Appendix–A

Schrödinger's Derivation of Wave Equation

Schrödinger put

$$S = K \ln \psi \tag{A.1}$$

in the Hamilton–Jacobi equation

$$H\left(q_j \cdot \frac{\partial S}{\partial q}\right) = E \tag{A.2}$$

where K has dimensions of action and ψ is a new unknown function. Substituting (A.2) in (A.1) gives

$$H\left(q, \frac{K}{\psi}\frac{\partial \psi}{\partial q}\right) = E. \tag{A.3}$$

Schrödinger proposed that H should be a quadratic form of ψ and its first derivative (ψ being everywhere real, single-valued functions) continuously differentiable up to second order. For the Kepler problem, Schrödinger chose

$$H = \frac{K^2}{2m\psi^2}\left[\left(\frac{\partial \psi}{\partial x}\right)^2 + \left(\frac{\partial \psi}{\partial y}\right)^2 + \left(\frac{\partial \psi}{\partial z}\right)^2\right] - \frac{e^2}{r} \tag{A.4}$$

so that

$$\left(\frac{\partial \psi}{\partial x}\right)^2 + \left(\frac{\partial \psi}{\partial y}\right)^2 + \left(\frac{\partial \psi}{\partial z}\right)^2 - \frac{2m}{K^2}\left(E + \frac{e^2}{r}\right)\psi^2 = 0. \tag{A.5}$$

The variation problem, he stated, is

$$\delta J = \delta \int d^3x \left[\left(\frac{\partial \psi}{\partial x}\right)^2 + \left(\frac{\partial \psi}{\partial y}\right)^2 + \left(\frac{\partial \psi}{\partial z}\right)^2 - \frac{2m}{K^2}\left(E + \frac{e^2}{r}\right)\psi^2\right] = 0 \tag{A.6}$$

from which he got, taking $K = \hbar$,[†]

$$\nabla^2 \psi + \frac{2m}{\hbar^2}\left(E + \frac{e^2}{r}\right)\psi = 0, \tag{A.7}$$

[†]It may be noted here that in to-day's quantum mechanics K is taken to be $(-i\hbar)$ and ψ as complex rather than real. The prescription $\vec{p} = -i\hbar\vec{\nabla}$ follows from (A.1) $\vec{p} = \frac{\partial S}{\partial \vec{p}} = \frac{\imath\hbar}{\psi} \cdot \frac{\partial \psi}{\partial \vec{q}}$ which is $\vec{p}\psi = -i\hbar\frac{\partial \psi}{\partial \vec{q}}$.

$$\int dS_n \delta\psi \frac{\partial \psi}{\partial n} = 0, \tag{A.8}$$

where the surface integral (A.8) serves as a boundary condition.

Appendix–B

Eigenfunction for Azimuthal Action Operator

Here we give the derivation of formulae (9.8a) and (9.8b) for the eigenfunction of the azimuthal action operator $S_a = \vec{\sigma}.\vec{L}$ corresponding to its eigenvalues l and $-(l+1)$.

Since the operator S_a is a 2×2 matrix operator, its eigenfunction must be a two-component spinor. Since it contains $\vec{L}$ as a factor, each component will be a spherical harmonic. Therefore, we have

$$S_a \Omega_{lm} = S_a \begin{pmatrix} f & Y_{l,m}(\theta, \varphi) \\ g & Y_{l,m'}(\theta, \varphi) \end{pmatrix} = l \begin{pmatrix} Y_{l,m}(\theta, \varphi) \\ Y_{l,m'}(\theta, \varphi) \end{pmatrix} \tag{B.1}$$

with

$$S_a = \vec{\sigma}.\vec{L} = \sigma^+ L^- + \sigma^- L^+ + \sigma_3 \; L_3, \tag{B.2}$$

where

$$\sigma^{(\pm)} = \frac{1}{2}(\sigma_1 \pm i\sigma_2), \quad L^{\pm} = L_1 \pm iL_2. \tag{B.3}$$

Using standard operations, equation (B.1) takes the form

$$f(m-l)Y_{l,m}(\theta, \varphi) + g\sqrt{(l+m')(l'-m'+1)'}\, Y_{l,m'-1}(\theta, \varphi) = 0, \tag{B.4}$$

$$g(l+m')Y_{l,m'}(\theta, \varphi) + f\sqrt{(l+m+1)(l-m)'}\, Y_{l,m+1}(\theta, \varphi) = 0.$$

Solutions of these two coupled equations is easily seen to be

$$m' = m+1, \; g - f\sqrt{\frac{l-m}{l+m+1}}, \tag{B.5}$$

which together with normalisation condition

$$f^2 + g^2 = 1 \tag{B.6}$$

gives

$$f = \sqrt{\frac{l+m+1}{2l+1}}, \quad g = \sqrt{\frac{l-m}{2l+1}} \tag{B.7}$$

so that we have

$$\Omega_{lm}(\theta,\varphi) = \begin{pmatrix} \sqrt{\frac{l+m+1}{2l+1}} & Y_{l,m}(\theta,\varphi) \\ \sqrt{\frac{l-m}{2l+1}} & Y_{l,m+1}(\theta,\varphi) \end{pmatrix}. \tag{B.8}$$

Formula (9.8b) can be derived in the same manner.

Subject Index

Author Index